COLLECTED WORKS OF
CHARLOTTE WOLFF

Volume 3

THE HAND IN PSYCHOLOGICAL DIAGNOSIS

THE HAND IN PSYCHOLOGICAL DIAGNOSIS

CHARLOTTE WOLFF

Routledge
Taylor & Francis Group
LONDON AND NEW YORK

First published in 1951 by Methuen & Co. Ltd.

This edition first published in 2016
by Routledge
27 Church Road, Hove BN3 2FA

and by Routledge
711 Third Avenue, New York, NY 10017

Routledge is an imprint of the Taylor & Francis Group, an informa business

British Library Cataloguing in Publication Data
A catalogue record for this book is available from the British Library

ISBN: 978-1-138-92367-6 (Set)
ISBN: 978-1-315-67907-5 (Set) (ebk)
ISBN: 978-1-138-93195-4 (Volume 3) (hbk)
ISBN: 978-1-315-67911-2 (Volume 3) (ebk)

Publisher's Note
The publisher has gone to great lengths to ensure the quality of this reprint but points out that some imperfections in the original copies may be apparent.

Disclaimer
The publisher has made every effort to trace copyright holders and would welcome correspondence from those they have been unable to trace.

THE HAND IN PSYCHOLOGICAL DIAGNOSIS

by

DR. CHARLOTTE WOLFF

FELLOW OF THE BRITISH PSYCHOLOGICAL SOCIETY

METHUEN & CO. LTD. LONDON

36 Essex Street, Strand, W.C.2.

First published in 1951

CATALOGUE NO. 5380/U

PRINTED IN GREAT BRITAIN

To
all those who have contributed
directly or indirectly to
this book's existence

CONTENTS

ILLUSTRATIONS AT PAGE 210

HANDS OF MENTAL DEFECTIVES

HANDS OF HIGH-GRADE MENTALLY DEFECTIVE AND OF NORMAL CHILDREN

PHOTOGRAPHS OF THE PRINCIPAL HAND-TYPES

INTRODUCTION

I HAD felt uncertain whether to publish another book on the human hand now because it is not yet possible to give an all-round picture of all aspects of the subject. I am convinced, however, that enough evidence has already been collected through my clinical studies to communicate with confidence and certainty a number of results which can serve as means of psychological diagnosis.

The study of the hand in relation to psychological diagnosis, as I am pursuing it, is a new branch of psychology. It must not, therefore, be regarded or judged from the point of view of perfection or completeness. As is always the case in a pioneer study, my work is in process of development and my views undergo changes according to my findings.

Practical experience and a deeper understanding of psycho-motor phenomena have altered my theoretical views in many respects during recent years, and the reader will find in this book a number of deviations from ideas expressed in *The Human Hand*, and I reserve to myself the right to alter and extend what I have written now in my next book.

The second part of this book is based on the many research studies I have undertaken since I wrote *The Human Hand*. I have studied the Hand of the Mental Defective on a large scale, with a special view towards the constitutional factor of moral deficiency. To assess the latter I paid several visits to a Colony for Mentally Defective Children, where I was able to investigate the hands of 275 high-grade defective boys and girls, many of whom had already been before the magistrate.

I have studied the hands of normal boys and girls in several L.C.C.- and other schools, in order to compare those of the mentally defective and the normally intelligent child. Such a study is not only important for the advance of psychological knowledge, but has great practical implications of which I shall speak further on in this book.

I have also investigated the hand in mental illness on a large scale, with a view towards understanding both the 'abnormal' and the 'normal' personality, as far as the latter exists. It is a banality to mention that we only know about the 'normal' person by knowing about the abnormal one.

Perhaps the most important study I have undertaken in recent years is that of the endocrine factor as seen in the hand. I had the good fortune to work at the Endocrine Clinic of Dr. S. L. Simpson at Willesden General Hospital where I saw a good number of practically all the major endocrine syndromes. The endocrine factor is a keypoint for understanding the link between physiological function and psychological reactions, and the diagnosis of the endocrine make-up automatically yields a diagnosis of certain features of temperament.

The chapter on the Endocrine Factor has thus important theoretical implications and for this reason I have put it into the first part of my book, in spite of the fact that this study was a clinical one.

I would never have been able to write this book without the help of all those who have given me the opportunity for clinical study. I owe a very great debt of gratitude to them. Professor Wallon, France, was the first to encourage my research and to give me the opportunity to assist at his consultations, and to visit the Home for mentally defective children at Livry Gargan, of which he is the supervising psychiatrist. He not only opened the door to clinical research to me, but he also opened my eyes to the importance of psycho-motor reactions in both the psychological significance of gesture and the origin of certain hand-traits.

I also want to express my gratitude to Dr. Stephenson, who supervised my researches at University College, London, and to Professor Sir Cyril Burt, who accepted me as a research student there.

I owe a special debt of gratitude to Dr. C. J. C. Earl, who has helped and advised me for many years. My studies on 'The Hand of the Mental Defective' at St. Lawrence's Hospital, Caterham, were made under his supervision. Since then I have visited the Institution, of which he is now the Medical Superintendent, at several intervals, and my study, 'The Form and Dermatoglyphics of the Hands of 115 Difficult and High-Grade Boys' was carried out there. My research at this Colony is by no means terminated, but is to be continued.

I have been most fortunate to be able to compare the hands of mental defectives with those of normal children through my work at several schools. I am especially indebted for this research to the Heads of the Schools concerned and to their teaching staff, particularly to Miss Lucy Preston, who made some valuable suggestions and helped me in other ways as well.

I also want to thank Dr. J. L. A. Lewis, former Superintendent of St. Bernard's Hospital, Southall, and Dr. T. Lindsay, former

Superintendent of St. Lawrence's Hospital, Caterham, for allowing me to study at their respective hospitals, as well as Dr. S. L. Simpson for giving me the opportunity for endocrine research. I cannot name the many others who have actively helped my work, but I am grateful to them all.

Apart from the direct help I have received, I owe much to some of the great pioneers in the study of the hand, such as Dr. Carus, the German physician of the King of Saxony in the middle of the nineteenth century, and Dr. Vaschide, who was Assistant at the École des Hautes Études, Paris; last but not least to Sir Charles Bell, who wrote the illuminating book *The Hand: Its Mechanism and Vital Endowments, an Evincing Design.*

I regret that I omitted in *The Human Hand* to pay tribute to this great man, who saw the whole psychological significance of the hand a century ago. His ideas and interpretations are still valid in the light of modern research, and I would like to quote some passages from his book which are particularly relevant to my subject.

The following passage is to be found in Chapter X of Sir Charles Bell's book (p. 252):

> Seeing the perfection of the human hand, both in structure and endowments, we can hardly be surprised at some philosophers entertaining the opinion of Anaxagoras, that the superiority of man is owing to his hand. Although the system of bones, muscles and nerves which belongs to this extremity, is suited to every form and condition of vertebrated animals, yet it is in the human hand that we perceive the consummation of all perfection as an instrument. This superiority consists in its combination of strength, with variety, extent, and rapidity of motion; in the power of the thumb, and the forms, relations, and sensibility of the fingers, which adapt it for holding, pulling, spinning, weaving, and constructing; properties which may be found separately in other animals, but are combined in the human hand.
>
> In virtue of these provisions, the hand corresponds to the superior mental capacities with which man is endowed. The instrument is capable of executing whatever his ingenuity suggests.

On the evolutionary significance of the hand, Sir Charles Bell says this (p. 272):

> Our argument, in the early part of the volume, has shown Man, by the power of the hand AS THE READY INSTRUMENT OF THE MIND accommodated to every condition through which his destinies are to be accomplished. We first see the hand ministering to his necessities, and sustaining the life of the individual. In a second stage of his progress, we see it adapted

to the wants of society, when man becomes a labourer and an artificer. In a state still more advanced, science is brought in aid of mechanical ingenuity, and the elements which seemed adverse to the progress of society, become the means conducing to it.

About the significance of the sense of touch localized in the hand he says:

We find every organ of sense, with the exception of that of touch, more perfect in brutes than in man. In the eagle, hawk, gazelle, and feline tribe, the perfection of the eye is admirable; in the dog, wolf, hyaena, as well as birds of prey, the sense of smell is inconceivably acute; and if we hesitate to assign a more exquisite sense of taste to the inferior animals, we cannot doubt their superiority in that of hearing. But in the sense of touch, seated in the hand man claims the superiority; and it is of consequence to our conclusion that we should observe why it is so.

And also (p. 224):

The extremities of the fingers best exhibit the provisions for the exercise of the sense. The nails give support to the tips of the fingers; and in order to sustain the elastic cushion which forms their extremity, they are made broad and shield-like. This cushion is an important part of the exterior apparatus; its fullness and elasticity adapt it admirably for touch. An ingenious gentleman has observed that we cannot feel the pulse at the wrist with the tongue. That is a remarkable fact; and I apprehend that it is owing, not to the insensibility of the tongue, but to its softness of texture; the tip of the tongue is not fitted to receive that peculiar impulse, to which the firm and elastic pad of the finger is so perfectly suited. Is it not interesting to find, that had the organ of touch been formed as delicately as the tongue, we should have lost one of OUR INLETS TO THE KNOWLEDGE OF MATTER.

About the muscular power of the fingers, especially that of the thumb, Sir Charles says (p. 114):

The thumb is called pollex, because of its strength; and that strength being equal to that of all the fingers, is necessary to the perfection of the hand. Without the fleshy ball of the thumb, the power of the fingers would avail nothing; and accordingly the large ball formed by the muscles of the thumb, is the distinguishing character of the human hand, and especially of that of an expert workman. The loss of the thumb amounts almost to the loss of the hand; and were it to happen in both hands, it would reduce a man to a miserable dependence: or as Adoni-Bezek said of the threescore and ten kings, the thumbs of whose hands and of whose feet he had cut off, 'they gather their meat under my table'.

I will end by quoting two more passages, the first concerning the link between the hand and intelligence, and the second the relation between the hand and spatial conception (p. 40 and p. 234):

> But as we proceed, we shall find how the Hand supplies all instruments, and by its correspondence with the intellect, gives him universal dominion. It presents the last proof in the order of creation, of that principle of adaptation which evinces design.

and

> Gradually, the length of the arm, and the extent of its motions, become the measure of distance, of form, of relation and perhaps of time.

The insight and knowledge of Sir Charles Bell, who saw the psychological significance and importance of the hand so clearly, makes one wonder why the study of the hand as a means of psychological diagnosis has never been attempted by scientists. Even the authority of this great man could not clear the subject of hand-interpretation from the taboo of charlatanism, into which it had fallen since the earliest times.

This book, *The Hand in Psychological Diagnosis*, is designed to contribute some new diagnostic possibilities to psychology and psychiatry.

CHAPTER I

THE HAND AND INTELLIGENCE

THE signal change in human evolution was due to the upright gait. When Man lifted his hands from the ground they were freed from the purpose of locomotion and were put into the service of intelligence. When the era of Instinctive Man ended, the brain developed new functions which created the human mind. That the use of the hand is intrinsically linked with intelligence, both phylogenetically and ontogenetically, has been stressed by philosophers since antiquity. Anaxagoras dwelt upon this problem—whether it was primarily the use of the hand which developed a human intelligence superior to that of the animal—while Aristotle wondered if the superior intelligence of man led to the development of new functions of the hand.

Instinct is also a form of intelligence, but what essentially distinguishes the two is the unconscious working of the former, and the conscious direction of the latter. Something which is done by instinct has a completely different quality from what is done by conscious impulses even when the actions themselves look alike. Conscious impulses lead to choice and discrimination in regard to both the goal to be achieved and the means to achieve it. Discrimination goes with differentiation in both the mental and the motor sphere. The differentiation of movements develops manual skill, which cannot be divorced from intelligence. It is, however, possible to possess intelligence without manual skill.

In order to clarify seemingly different forms of intelligence, a definition is necessary. Intelligence can be defined as the capacity to learn and to acquire knowledge. The speed of learning is in direct proportion to the degree of intelligence. Speed of learning by itself leads to superficial knowledge only, if it is not accompanied by the capacity of mental assimilation, which represents some form of a creative process of the mind. The scope of intelligence must also be measured according to the range of subject matters which are within a person's grasp. The speed of acquiring knowledge, the power to assimilate it, and the variety of subject matters determine the level of intelligence. Apart from this, one must differentiate a more concrete from a more abstract form of intelligence. The prevalence of the one over the other is already recognizable at a fairly early age.

In people with manual skill one generally finds a better-developed concrete intelligence than in people who are 'gauches'. The latter often possess, on the other hand, a highly developed abstract or symbolic intelligence.

It was H. Wallon who pointed out in his writings and lectures how intimately intelligence is linked with motor function in general and with that of the hands in particular. One has only to enter the wards of a home for mental defectives to be struck by the awkwardness of their movements and their lack of motor co-ordination. Undoubtedly mental deficiency goes with motor deficiency. But if that is so, why does one find in a great number of highly intelligent and cultured people gaucherie and sometimes even lack of motor co-ordination? There are many among them who do not know how to hold a tool properly in their hands, still less have they the capacity for using it. To understand this contradiction one must remember that intelligence is a function of the mind, and not a water-tight compartment, and that it is interrelated with the imagination as well as with emotional reactions. It is from this background that the natural flow of psycho-motor reactions can be put out of gear, with resulting motor inhibitions. As I have already said, such a failing is found in connexion with the abstract type of intelligence, which is characteristic of the predominantly 'introvert' personality. Introversion goes with marked inhibitions both in emotional reactions and motor impulses. This is easily understandable through the following consideration: in 'introverted' people concentration of thought is mainly turned inwards and focused on imaginative processes. Awareness of the outside world suffers, and owing to their absorption with their own inner world 'introverted' people perceive changing situations in their surroundings almost as minor shock experiences. However slight these may be, they cause their reactions to get out of tune, and this leads to abruptness and inhibition of their psycho-motor responses. The gaucherie of the intellectual can therefore be explained by considering features of both temperament and imagination. This shows how careful one has to be with the interpretation of intelligence tests, which may sometimes place an intelligent person along with a feeble-minded one. One has to be especially careful in judging intelligence from performance tests which demand a strong concrete sense and manual skill, and it can be said that with a number of people of a special make-up such tests are rather indicative of temperament than of intelligence. A good illustration of this state of affairs is a little episode reported in the *Daily Telegraph* of 28th May, 1948, under the heading of 'Subnormal Psychiatrist'. It reads:

> I was particularly amused when he [Lord Cherwell] ended his criticism of the use of psychiatry with the story of the distinguished psychiatrist who rashly submitted himself to one of the tests. He was marked: 'Definitely subnormal, should see a psychiatrist.'

It is common knowledge that excellence in some capacities often goes with deficiency in others. If such a discrepancy develops to the extreme, some form of disintegration of the personality is unavoidable. The constitutional disposition of the highly intelligent 'introvert' always shows a neurotic streak.

Now let us take the opposite type to that of the person who suffers from motor inhibition and lack of manual skill, namely, the person who possesses a perfect command over his movements: the dancer. Nobody can doubt that there are intelligent dancers, but as a rule intelligence is not the striking feature of their make-up. In them, skill and precision of movements as well as grace and elegance come from the instinctive and imaginative sphere, rather than from cortical impulses. Perfect psycho-motor expression can go with a low intelligence. The dancer belongs, as a rule, to the kinaesthetic type of person, with the characteristic narcissistic attitude of behaviour and emotional immaturity. The motor skill of the dancer must not be confused with motor precision which is cortically controlled. In the kinaesthetic type, motor images prevail over visual and phonetic images and postural sensitivity is unusually high. It is therefore through a natural knack of his psycho-physical constitution that the dancer achieves in the first place his kinaesthetic art. Unquestionably he has to work hard and to use his intelligence, but the subcortical impulses are the essential background from which his performance is directed.

A highly skilled craftsman, whose hand movements are as selective as they are precise must, on the contrary, have intelligence, even if only in a specialized form. And, as a rule, his intelligence is of a concrete type. The artisan is, in fact, the best-balanced human being in regard to the unison of cortical impulses and hand movements, and so is the sportsman, especially in those sports where the use of the hands takes first place. The perfect control of the motor apparatus is, however, only one small section of the vast field of cerebral activities. Both the artisan and the sportsman are representatives of a 'departmental' perfection. Small wonder that the latter especially was selected as the desirable type of human being by certain peoples or 'races', not to their cultural advantage. Practical intelligence together with motor skill are by themselves a very small part of the human mind, and only when they are accompanied by a rich emotional life, a colourful imagination and a rich sensitivity are they the mark of a highly developed personality.

This explanation will make it clear how complex and difficult the assessment of intelligence really is. It seems to me to be essential to establish a test which will render justice to both the different forms of intelligence, and the specific individual factors. Intelligence cannot be taken as absolute but must be considered as a part of the whole personality. Many intelligence tests have been devised for both normal and psychotic individuals, but all of them present the same handicap in that they produce a situation to which most people react emotionally, either with anxiety or with a certain exuberance due to an exhibitionistic streak. Such reactions must falsify results. It is obvious that an intelligence test which is objective, and which could be applied without the person being aware of it, would be the ideal test. Can the hand serve such a purpose? I must make it clear from the start that the hand does not represent the means of a quantitative estimate of intelligence, and cannot, therefore, replace any of the intelligence tests in use but can, as I am able to demonstrate, compensate for the shortcomings which I have just described.

The hand is an organ of touch as well as of movement, and it serves also as a vehicle of expression through gestures. I have stressed the significance of gestures as a conveyance of subconscious impulses in my book *A Psychology of Gesture*. In its role as the principal organ of gesture the hand is linked more with the emotional Self than with intelligence. The hand as an organ of touch as well as of movement is one of the most important tools of learning, and is thus linked with the development of intelligence. The way of touching objects and learning about their shape and other properties is strictly individual and must leave in one way or another a mark on both the mobility of the hand and fingers and their shape. Repetitive movements of the fingers in their individual variety necessarily involve the development of crease-lines, particularly in the upper half of the palm. These can therefore be considered in one way or another to be linked with the intellect.

I have pointed out before that intelligence is a function of the mind which is interrelated with the whole of the personality. The subtle emotions akin to feeling are part of the intellectual outfit and even more so is the imagination, which cannot be separated from intelligence. The power of assimilation of knowledge largely depends on the quality of the imagination. The latter plays also a tremendous part in emotional as well as sensitive reactions, both of which express themselves in gestures and postures of the hand. One may call the imagination the BRIDGE BETWEEN EMOTION AND INTELLIGENCE. However abstract and categorical our thinking may be, it is built on imagery. The more the intellectual process becomes rationalized and logical, the more mental images tend to lose their original freshness and to become mere symbols comparable with shorthand writing. Through such a simplification the process of thinking is at once quickened and generalized. But in the same way as shorthand develops from the complex symbols of handwriting, which in their turn go back to pictorial writing, abstract thought has its roots in mental images. These originate from perceptions of the outside world, from objects seen and felt and situations experienced. There can be nothing in our mind which has not at one time or another been experienced by our senses. The senses which are the most important for building up our stock of knowledge are the sense of touch and the visual sense. The former is most intensified and differentiated in the hand. While visual perception assures pictorial representation in the cortex of the brain, the sense of touch and the kinaesthetic sense assure the spatial orientation of images. Without the latter mental images would be without perspective.

The hand as both the tool of learning about the outside world and as an organ of spatial sensibility can be considered the fundamental vehicle of the structure of thought. Spatial differentiation goes with physical as well as mental 'balance', and gives to perception the tridimensional aspect essential to the orientation of man in the outside world and to its representation in his mind. One can therefore say that the hand as an organ of sensitivity assures spatial or tridimensional imprints of the outer world in the cortex of the brain. This is of paramount importance for the development of intelligence, because without their spatial character mental images would not only be flat, but they could not be remembered. The faculty of memory is bound up with spatial orientation and a structural sense, which are both essential to bring order into the whole material of images and associations. The mind would be a chaos without spatial orientation. As everybody will know from his own experience, the clarity of images helps in memorizing them.

The perception of objects goes automatically with psycho-motor reactions. This is especially marked in the child in whom the inhibitive factor is practically non-existent. Touching objects with one's hands involves automatically many movements of the hand. Their variety, quality and co-ordination must in the end reflect on mental imagery, in particular on the clarity and spatial discrimination of the images and also on the categorical order of thought. This might explain the phenomenon of gesticulation which is nothing else but a spatial expression of images and thought. Especially marked is this 'drawing' of thought in space in creative and original thinking, when the mind is fighting for clarity. H. Wallon has pointed out in his *L'Enfant Turbulent* that all mental imagery is accompanied by more or less slight subconscious gestures of the hand, and even when the latter appears to be still during the process of thinking and day-dreaming, it is always held in certain postures which change according to the feeling-tone of the images. Our mind goes on working not only when we are awake, but also when we are asleep, especially on ideas or problems which are puzzling. A friend of mine told me how surprised she was when she saw that her small son, aged 7, blissfully asleep, continually moved his right hand in a peculiar fashion. She suddenly understood what these movements meant: they were gestures of milking. The boy had been in the country and had been intrigued by the milking of a cow. He had badly wanted to have a try himself, but for obvious reasons was not allowed to do so. The mental process of milking went on and on in his mind and expressed itself in movement as a wish-fulfilment in his dream. It is safe to say that there is practically no mental activity in which the hand is not in one way or another involved.

The link between the hand and intelligence is obvious, but how can the latter be tested from hand-features? It is clear from what I have explained that the shape and other characteristics of the fingers as well as the crease-lines, which are due to finger movements, can indicate intelligence. This is also borne out by a negative deduction, namely through a marked difference in the shape and size of the fingers in mental defectives and intelligent people. In the clinical studies in this book these differences are demonstrated in detail. The development of the nails is correlated with the shape of the fingers, a fact which I have already stressed in my book *The Human Hand*. Again, deviations in the shape of the nails from the normal in mental deficiency are a practical proof of their significance in regard to intelligence.

Two fingers occupy a special place in the evolutionary scale: the thumb and the index finger. The mobility of both is unique in Man. With monkeys and apes the forefinger cannot be used separately. The thumb, particularly in anthropoid apes, is minimal, and its mobility is very restricted. One can therefore assume that the special development of the two fingers in Man is correlated with the development of human intelligence. The freedom of movement of the thumb and index finger leads to the development of certain crease-lines, both principal and accessory ones, the properties of which are likely to be indicative of intelligence also.

The index finger, as its name implies, is an indicator. It either points at or it points out something. This use of the forefinger transmits the human faculty of discrimination, which is an essential quality of intelligence. The index finger as well as the thumb is also used to emphasize; while the former generally underlines contents of thought, the latter emphasizes the Ego of the speaker. A person who makes habitually pointed movements of the thumb is certainly of an aggressive and self-assertive nature. People who frequently use the index finger in a pointing gesture are inclined to be pedantic and self-righteous. But apart from emotional impulses which cause certain gestures of the two first fingers, they are constantly used in manual activities, both actively—in movement, and passively—in touching objects, or, in other words, in 'feeling' things. That is why thought and feeling are particularly linked with these two fingers. As I have already said, the index finger cannot be moved separately in apes and monkeys, a fact which is reflected in their fourfinger line, which runs straight from one edge of the palm to the other. In human beings there are as a rule two fourfinger lines which run in a curved fashion. In their 'élan' and other qualities of design these two lines are likely to be correlated with intelligence. It is not only the two main transverse creases which are indicative in one way or another of intelligence, but also the quality and design of all creases due to the movements of the first two fingers.

Now, it can rightly be argued that all hand movements have two different sources: they either come from cortical or subcortical (emotional) impulses. How is it then possible to differentiate from the formation of the fingers and the creases qualities of intelligence from those of temperament? I have to disappoint those readers who expect an easy answer, a yes or a no in psychological matters. There is no strictly dividing line between temperament and intelligence, and there is no separate indication of either in hand-features. Intelligence, imagination, feeling and emotion are a psychical entity and from this wholeness only can one come to conclusions about the character and quality of the parts. One can even go so far as to say that a certain temperament goes with a certain type of intelligence, and in knowing something about the former one automatically knows something about the latter. This means that intelligence can only be indirectly assessed by means of hand-features, and this precludes any quantitative estimate. We know through the work of Kretschmer and Sheldon that constitutional dispositions are interrelated with each other. For example, the 'cyclothyme' personality (Kretschmer) and the man with predominant endomorph and mesomorph factors (Sheldon) possess as a rule a more concrete and practical intelligence than the 'schyzothyme' personality (Kretschmer), whose thinking is, on the whole, more on symbolic lines.

Let us take some examples in order to show how one can estimate intelligence from hand-features.

EXAMPLE I

Photograph 5 and print 4 represent an example of the small sensitive hand of the broad variety. These are well-proportioned hands, the fingers are strong but very mobile, with tips of medium length and well-developed nails which have a square shape. The finger-tips are flexible. The bumps of sensitivity are strongly marked, especially on the fourth and fifth fingers. The thenar and hypothenar eminences are strong and the latter is very deeply inserted, as is quite often the case in this type of hand. There is a clear composition of the crease-lines, a harmonious balance between the well-designed four main creases and the accessory ones, which are most clearly and deeply marked in the distal part of the palm. This hand goes with a highly nervous and sensitive temperament and a personality of great emotional and intellectual richness. The imaginative factor plays a decisive part in both the sphere of temperament and the sphere of intelligence. This is the hand of a gifted, versatile personality with a great power of expression. The latter is indicated by both the shape of the fingers and the characteristics of the accessory crease-lines in the distal part of the palm. The temperament is characterized by a good mixture of extravert and introvert tendencies—awareness and sensitivity towards the outer world are the marks of the former, a highly developed imagination the mark of the latter.

The intelligence can be described as follows: A quick grasp of any problem which concerns the emotional and imaginative sphere;

the capacity of abstract thought not on philosophical, but on imaginative lines. This is much more the intellectual make-up of an artist than of a scientist. The practical and concrete sense are probably much inferior to the powers of imagination and abstract thought. Aesthetic understanding and taste are highly developed. There is a sensitive discrimination as well as capacity of judgement in both aesthetic and human matters.

EXAMPLE II

Now let us discuss photograph 4 and print 3 which represent a bony motor type of hand, by which is meant that the bony motor factor is predominant. Other factors, for example a sensitive element, shown in the emphasized hypothenar eminence and the long rather thin fingers with long square and tapered terminal phalanxes, are present. There is a multitude of accessory crease-lines, particularly on the distal part of the palm and on the hypothenar eminence. This hand goes with a predominantly 'introvert' personality, one that has a rich imagination, but too strong an inhibitive factor which interferes with the spontaneous expression of the emotions. These are conflicting and complex and lead to an inner restlessness, as can be seen by the multitude of creases and their hesitant and broken-up design. The shape of the fingers and the characteristics of the creases suggest a very receptive personality in whom both imaginative and abstract thinking far outweigh practical sense and concrete thinking. These hands belong to an imaginative, predominantly 'introvert' man with high intellectual capacity on abstract and literary lines. His intellectual potentialities are higher than his manifest intelligence, because of the amount and intensity of conflicting emotional tendencies.

EXAMPLE III

Print 2 represents the palmar surface of a well-known man, whose name I shall disclose at the end of this analysis. He has fleshy motor hands with markedly long fingers, the fifth of which is abnormally long. The finger-tips are rather characteristic of the 'sensitive' type of hand. The hands are unusually broad like that of a heavyweight. They look pugnacious, and were it not for the characteristics of the fingers and the beautiful verve and design of the creases, one would think that they were the hands of a boxer. The lower transverse crease and the long longitudinal line are of the most perfect design. The whole composition of the crease-lines is beautifully spaced; the lines in themselves recall those drawn by an artist; they are neither too heavy nor too superficial, and of a dynamic élan.

These hands belong to a personality with predominantly extravert tendencies. Introversion here is localized in the domain of thought and manifests itself in a tendency to contemplation in a more oriental fashion than is natural to an English temperament. The unusual physical strength and his mental as well as physical pugnaciousness make this man try his hand at many different things. His versatility does not go, as is so often the case, with a lack of concentration; he is completely absorbed in whatever he is doing. His intelligence is of a predominantly concrete and at the same time polemic kind. He pays great attention to detail but goes at the same time to the heart of the matter of any subject he is tackling. He ought to be a good speaker, especially in discussions in which there is something 'to fight for'. The hands belong to the writer H. N. Brailsford.

EXAMPLE IV

The last of my examples represents the hands of a feeble-minded person, in order to show the contrast in hand-features of the highly intelligent and the subnormal person. Print 6 has been taken from the hands of a boy of 13 years of age. It must be clear even to the untrained eye that the fingers, particularly the thumb, appear to be not only stiff but clumsy and somewhat immobile. They are, so to speak, 'inarticulate'. This lifelessness of the digits is accentuated by the poverty of the creases. There are a minimum of them. The two main transverse lines are joined into one Simian-like line. The long longitudinal line is very short and there are almost no accessory lines. Both the main creases and the few accessory ones are of a 'heavy' design. The palm has a normal width, but is very flat. This hand belongs to a primary type of personality with a dull schyzothyme temperament and a subnormal intelligence.

It is evident from the examples described that an indirect and qualitative estimate of intelligence is possible from hand-features. There can be no doubt that a trained person can diagnose at a glance:

(i) If the subject tested is of normal or subnormal intelligence.

(ii) If he or she is more inclined towards concrete and practical or towards abstract and symbolic thought.

The ordinary reader might argue that such a general diagnosis is not very admirable and helpful. The physician and psychologist will, however, realize that such a rough estimate, if reliable, is of very great value especially in border-line cases of normal and subnormal intelligence, when great harm can be done through a misinterpreted intelligence test. It is not an exceptional mistake to place a normally intelligent child along with a mental defective, because he appears to be backward. In fact, his backwardness is due to some kind of emotional disturbance. Apart from the scientific interest of intelligence diagnosis from hand-features, the educational and humanitarian value of a reliable test for border-line cases cannot be over-estimated.

CHAPTER II

THE ENDOCRINE FACTOR

I BEGAN systematic research on hand-traits sixteen years ago and have come to the conclusion that the endocrine factor is one of the key points in hand interpretation. Endocrine disturbances are known to be linked with psychoneurosis, psychosis and mental deficiency. It is, therefore, reasonable to assume that the endocrine glands are intrinsically linked with temperament and mentality, and that the diagnosis of the endocrine make-up of a person includes an indication about his or her emotional make-up. I am convinced that this is so, but many endocrinologists and physicians still deny such a psychosomatic conception. As everybody knows, endocrinology is still in its infancy and much of the complex interrelations of the functions of the various endocrine glands is still unknown. As both endocrinology and psychosomatic medicine are pioneer subjects it is understandable and wise that one should be careful in accepting quick results. In spite of this reservation, enough data are available to prove that close correlations exist between the endocrine system and hand-traits and between the endocrine glands and personality. It is still difficult today to do research work on endocrine glands, as only comparatively few cases come into the hands of the general practitioner and there are not many which are sent to endocrine clinics.

Luckily I was able fourteen years ago to study the correlations between endocrine deficiency and hand-traits while I was working at the Dispensaire pour les Enfants Difficiles et Arriérés of Dr. Gilbert Robin in Paris. Dr. Robin laid great stress on the endocrine factor from both the diagnostic and the therapeutic aspect. The cases at his surgery were, of course, limited in number and also in the variety of endocrine disorders. A similar situation presented itself when I studied the hands of mental defectives. I recognized the endocrine factors which are often connected with mental deficiency, but I lacked a sufficient variety of cases to enable me to get a more thorough knowledge of the correlation between the endocrine system and hand-features. I was therefore particularly pleased when I was given the opportunity for a research study at the Endocrine Clinic of Dr. S. L. Simpson at the Willesden General Hospital. It was there that I found for the first time a fairly large number of cases of almost all types of endocrine disturbances and arrived at a new comprehension of their complex significance. I am deeply indebted to Dr. Simpson, not only for the opportunity he gave me and for his interest in my particular study, but also for the trouble he took to explain to me theoretically and practically the cases under consideration. I took hand impressions in a small number of cases only, but made careful notes of the general shape as well as the detailed characteristics of each pair of hands investigated. These notes are sufficient to call up in my memory every trait of the hands studied. As this study was a clinical one I should, logically, have put this chapter into the second part of my book. I thought, however, that it would be much more advantageous for the reader and the construction of the book to reverse the order, as I am using my clinical knowledge from a psychological angle and not from a medical one. It also seemed to me to be essential to write about the endocrine glands as seen in hand-features in the beginning of my book, as they represent a key point in hand-diagnosis, and the chapters which follow could not be properly understood without an explanation and discussion of endocrine significance in psychology.

I have pointed out already that the link which exists between the personality and the endocrine glands is by no means an accepted fact. There are, however, some over-enthusiastic protagonists of such a correlation, among whom I will mention only L. Berman and his provocative book *The Glands Regulating Personality* (1921). While Berman was carried away by enthusiasm rather than by scientific inquiry he has, nevertheless, done something to show up an important psychosomatic problem which had hitherto remained unnoticed. This was twenty-seven years ago. In our days one finds a number of psychiatrists who recognize the relationship between endocrine glands and personality while general physicians and the endocrinologists themselves are, on the whole, very guarded on the subject, as I mentioned at the beginning of this chapter. Dr. S. L. Simpson, who is one of the most progressive endocrinologists, acknowledges the mental and emotional implications, and a general outline of personality traits connected with endocrine syndromes is to be found in his important book *Major Endocrine Disorders* (2nd edition, 1948). While there are some references to personality in endocrinological books, practically no work has been done on the correlation between the hand and the endocrine glands. There are, to my knowledge, only two endocrinologists who mention the hand in this connexion; they are August Werner in his book *Endocrinology*

(1939 and 2nd edition, 1942), and S. L. Simpson in his book quoted above. Werner has given the hand a special place as a guide to endocrine diagnosis, and Simpson mentions hand-features among other markings of the physical constitution which go with certain endocrine dispositions. Werner has established altogether nine endocrine hand-types, but I have found his classification faulty, and that is why I made an independent approach towards the subject while working under the supervision of Dr. Simpson.

THEORETICAL CONSIDERATIONS

The bones of the hand are long bones and the premature or delayed closure of the epiphyses necessarily reflects abnormal endocrine function. C. Benda (1939) mentions that ossification of the long bones as well as of the carpal bones of the hand is early in hypothyroidism; this leads to the small, broad hands found in cretins and also in mongolian imbeciles. S. L. Simpson (1948) gives a different explanation of the hypothyroid hand. In the chapter on cretinism, which is a condition of hypothyroidism, he mentions the spadelike short hands but maintains that there is a late closure of the epiphyses. He explains the small physique and the short hands of the cretin through an abnormal functioning of the pituitary gland which inhibits their normal growth. I have given this example of hypothyroidism as seen in the hand in order to show:

(i) the correlation between endocrine dysfunction and hand-traits,

(ii) the different interpretation of the same phenomenon by different endocrinologists.

Both Simpson and Werner mention the elongated hand with very long and slender fingers in eunuchism and eunochoidism as being due to delayed closure of the epiphyses, a condition which is clearly shown in the X-ray photographs of the hands of such endocrine cases.

The hand in the 'Babinski-Froehlich' syndrome is also well known and indisputable. It is a pudgy, childlike hand (also in the grown-ups of this type), with pointed fingers and sausage-shaped phalanxes. And every physician as well as many laymen are familiar with the particular hand-trait of mongolian imbeciles. It is hardly worth mentioning that the hand in acromegaly is abnormally long and broad, as the hand is one of the most significant symptoms in this endocrine condition.

Much less known is the type of hand which goes with hypopituitary infantilism. Simpson mentions the unusually long and slender fingers connected with this condition, and I have given great prominence to the description of the hand-features of 'infantile' boys in my paper on 'The Form and Dermatoglyphics of the Hands of 115 High-Grade and Difficult Boys'.

Dr. Simpson also describes in his book already quoted, the curious deformation of the bones of the hand in parathyroidism and the marked pigmentation of the back of the hand in Addison's disease. The latter cases are of little interest for my study but are mentioned to show the significance of hand-traits in endocrine disease.

One distinguishes in endocrine disorders those which are present at birth and which are therefore congenital, and those which occur during lifetime. Congenital endocrine disorders are due to prenatal influences and have a genetic significance. They are very often found in a number of members of the same family. There can be no doubt that all congenital endocrine conditions show themselves in the shape of the hand as well as in other hand-traits, and any dysfunctions which arise before the epiphyses of the hand bones are closed must logically be marked in the structure of the hand bones and fingers. Endocrine disturbances which are acquired in later life are not necessarily reflected in hand-traits. I have seen a number of cases of myxoedema with long and thin hands which are the opposite to hypothyroid hands. Acromegaly, on the other hand, always affects the shape of the fingers at whatever time of life it manifests itself. The length of the bones naturally remains unaffected when the condition occurs after puberty, but their width alters, because the bones thicken through the abnormal stimulus of growth produced by hyperpituitarism. In a number of cases of thyrotoxicosis one finds thinning of the phalanxes of the fingers together with general slimming. This is, however, not always so. Thyrotoxicosis often goes with a hypothyroid condition as well and then, naturally, the hand shows rather stout fingers and sausage-shaped phalanxes. In order to understand the extremely complex and seemingly contradictory functioning of certain endocrine glands, the reader with a medical and psychological education should turn to Dr. Simpson's book from which I have quoted in this chapter.

It must be kept clearly in mind that the working of the endocrine glands is a concerted one and that all of them are functionally interrelated. The pituitary gland has, however, the predominant influence over them all and, as Dr. Simpson says in his book, from which I quote: 'If one only knew about the pituitary gland one would know about the whole of endocrinology.' It is beyond the scope of this book to go into a detailed discussion of this mysterious gland, but the following points must be stressed: The anterior lobe of the pituitary gland produces

(i) gonadotrophic,
(ii) thyrotrophic,
(iii) adrenocorticotrophic,
(iv) and growth hormones.

There are other types of hormones as well which belong to the series of those secreted by the anterior lobe, but they are of no interest to my subject.

The gonadotrophic hormones influence, as the name indicates, the sexual function, the adrenocorticotrophic hormones exert an influence over both sexual behaviour and muscular strength and the thyrotrophic hormones regulate metabolism.

It is also important to mention that pituitary function is interrelated with hypothalamic function. The hypothalamus is the most important seat of emotional and expressive behaviour and possesses also a representation of the autonomic nervous system. The influence of this link on certain aspects of human behaviour and temperament cannot be overrated. This short excursion into the physiology of the pituitary gland demonstrates its functional predominance and its tremendous influence on emotional reactions and expression.

There remains one more theoretical aspect to be considered, namely, whether the results of endocrine pathology can be applied to ordinary people. I have shown in my book *A Psychology of Gesture* that the difference between the mentally deranged and the normal mind is only one of degree, and the same applies to the endocrine make-up. The great variety of human physique, which has been so clearly stressed by Sheldon in his book *Varieties of Human Physique*, can only be understood by the great variety of endocrine functioning in different individuals. There are other factors, such as the nutritive one, for example, which determine human form, but the endocrine make-up takes first place in shaping our appearance. One can therefore assume that there is in every human being a characteristic endocrine formula which goes with a special psycho-physical make-up and which reveals, as in abnormal cases, only to a minor degree the predominance of one or another gland. Such a predominance creates in every individual a characteristic endocrine syndrome without involving, in so-called normal people, the presence of any clinical symptoms. The endocrine make-up must necessarily reflect on hand-features, as the bony structure as well as the shape of the hand and fingers are largely determined by endocrine functions.

ENDOCRINE HANDS

According to the predominance of either the pituitary or the thyroid, the gonads or the adrenals one can differentiate four principal endocrine hand-groups. The complex function of the pituitary gland is shown in three different hand-types belonging to the first group, and the same holds good for the thyroid hands.

The following tabulation shows the principal endocrine hand-types:

A. PITUITARY HAND-TYPES

1. THE HYPERPITUITARY HAND
2. THE PITUITARY INFANTILE HAND: CONGENITAL CONDITION
3. THE HAND IN PITUITARY OBESITY (FROEHLICH'S SYNDROME): CONGENITAL CONDITION

B. THYROID HAND-TYPES

1. THE CRETINOID HAND: CONGENITAL CONDITION
2. THE MYXOEDEMATIC HAND: CAN BE CONGENITAL, BUT IS NOT NECESSARILY SO
3. THE HYPERTHYROID HAND: NOT CONGENITAL

C. THE EUNOCHOID HAND: CONGENITAL

D. THE ADRENAL HAND

A. PITUITARY HAND-TYPES

1. THE HYPERPITUITARY HAND

This is an oversized hand. Pierre Marie described in 1897 'la main en large' and 'la main en longue' as indicative of hyperpituitarism. If the condition of hyperpituitarism occurs before the epiphyses of the bones of the hands are closed one finds *la main en longue*, if it arises at a later stage *la main en large*.

The Shape of the Hyperpituitary Hand

The hand is of an abnormal length and in many cases of abnormal width. The latter characteristic is particularly marked in cases of postpubescent hyperpituitarism. The fingers are very long and at the same time stout.

The Nails. The nails are usually of medium length and particularly broad, with or without moons. I have observed very large moons in 80 per cent of the cases I have seen.

Physical Qualities. The rigidity of the palm and the fingers is the outstanding feature. Other physical qualities differ widely according to the individual temperamental and physical make-up.

TEMPERAMENT AND MENTALITY

Hyperpituitary people are very moody with a tendency to a quick temper and depression. They are irritable and excitable. Obsessional or paranoiac tendencies are frequently found. One young man, whose hands I examined, showed features of violence, especially in his sexual life, which led him into crime. Hyper-excitability and hyper-sexuality are often symptoms at the onset of the illness, and give way at a later stage to impotence and loss of interest in life. While one finds hypomania associated in the beginning of acromegaly, depression is the prevailing mood of the mind later on.

The temperament presents the emotional curve which one finds associated with manic-depressive psychosis. Ideas of grandeur together with aggressiveness enhance a lack of adaptive powers which seem to be characteristic of this type of personality. There are very few mental defectives among hyperpituitary people. I only remember one case in the large number of mental defectives I have seen. But I know, on the other hand, three patients with acromegaly who were highly intelligent; one of them was a brilliant mathematician and chess player.

2. THE INFANTILE HAND (HYPOPITUITARY)

While hyperpituitarism is found comparatively rarely, pituitary infantilism is of a much more frequent occurrence and the infantile hand is one of those I have come across most often in both institutions for mental deficiency and in ordinary life. Dr. S. L. Simpson distinguishes two types of infantilism:

(i) the ideopathic or simple infantilism (congenital)
(ii) the hypothalamic infantilism, which is the result of severe shock or concussion, or infectious disease.

In all these cases the hypothalamic pituitary mechanism is inhibited.

Hypopituitary infantilism occurs either in early youth or in puberty before the epiphyses of the bones are closed. A special form of this condition, namely infantilism due to chronic infection, such as tuberculosis, had been recognized long ago by Levy Lorraine, but the syndrome he has described as 'Levy Lorraine infantilism' does not cover the whole range of causes and symptoms which we now understand by this term.

The infantile person is naturally both undersized and immature. In boys there is a feminine 'note', with lack of the secondary characteristics of the male sex. They need not shave and there is absence of the axillar and pubic hair. The 'infantile' person seems to carry eternal youth with him.

The Shape of the Infantile Hand (Group *a* : typical of people of normal intelligence):

The shape of the infantile hand is by no means the same in all cases. One must, therefore, speak of different groups of infantile hands which are characterized by different markings. In the majority of cases the infantile hand is well covered by the description of the sensitive small hand in my book *The Human Hand*. It is a small but well-proportioned hand, with short fingers furnished with either square or tapered tips and well-developed bumps of sensibility. The fingers are 'delicate' without being weak.

The Nails. The nails are either of medium size or short, and in the former case mostly have well-developed moons, which are missing in the latter case.

Physical Qualities. The palms are mostly hyperflexible and there is often hyperflexibility of the fingertips; the latter goes by no means always with the former. Other physical qualities vary too much to allow for a generalization. The skin is always of a delicate texture and is in many cases covered with down. There are numerous and fine (superficial) creases because of the delicacy of the skin.

The Shape of the Infantile Hand (Group *b*: typical of subnormal people):

In my paper 'The Form and Dermatoglyphs of the Hands of 115 Difficult and High-Grade Boys' (1947), I described a type of hand which was strikingly common among these boys and which occurred together with psycho-physical infantilism. The hand form of the subnormal infantile person differs widely from that which I have just described. The hand is much longer than the small sensitive type of hand, and this is especially so in the case of the fingers. The palm is generally of medium length with strongly developed thenar- and hypothenar eminences, while the distal part is strikingly flat and narrow. The fingers are very long and slender but strong. Spatulated terminal phalanxes are the rule, but square finger-tips also occur.

This very striking type of hand rather resembles at first glance the bony motor or leptosome hand. The remarkable length of the fingers might be due to secondary hypogonadic function.

The Nails. The nails vary widely in this type of hand, but as is often the case with mental defectives, they are mostly under-developed, very short and broadish and without moons.

Physical Qualities. The hands are in practically all cases rigid and in most cases cold and moist.

The Creases. I am giving a general account of the creases, as they are practically of a pattern and show definite characteristics. They are always few in number, especially on the distal part of the palm. They are deeply engraved and of a 'clumsy' design, as if drawn by the hesitant and 'gauche' hand of a child. Faults in the crease-lines occur in practically every case and the simian line is unusually frequent. It was present in 30 per cent of the hands of the 115 boys I described in my paper.

The Papillary Ridges. The papillary ridges as well as the creases show typical markings and are therefore included in the general description of hand-traits. Primary triangular patterns are frequent on the finger-tips and whorls and loops on the thenar and hypo-thenar eminences occur in the majority of cases. A statistical survey of the hand-traits of these boys can be found in the second part of this book.

TEMPERAMENT AND MENTALITY (Group *a*)

The sensitive small or infantile hand which occurs in people of either normal or above normal intelligence, goes with a nervous temperament and a neurasthenic streak. The emotional infantilism of the owner of this type of hand precludes maturity and the capacity to shoulder fully adult responsibilities. That is why one finds the infantile person attached to a stronger personality, either of the 'father' or 'mother' type. Nervous as well as neurotic disturbances are frequent. People with small, sensitive hands are generally endowed with a lot of charm and possess good manners. They desire to please. They use their charm to get what they want and need, which is firstly security, and secondly admiration, to bolster up their weak ego.

The infantile person of this group is versatile and gifted in many directions, but is lacking in depth. Superficiality of interest leads to instability and often to failure in professional activities, in spite of a certain brilliancy. Speed of comprehension and cleverness are the outstanding traits of the mental make-up.

TEMPERAMENT AND MENTALITY (Group *b*)

As I have already mentioned, this type of hand has been found in high-grade defectives mostly with asocial tendencies. I must recall here an instance which occurred at Dr. Simpson's Endocrine Clinic, and which illustrates best the personality connected with this type of hand. One day four young men, between the ages of 16 and 18, walked into the consulting-room accompanied by their teacher. All of them were undersized, decisively feminine in appearance and had a charming smile on their rather rough but handsome faces. They had been brought to the Clinic from a prison where they were serving sentences. They were unstable and nervous boys and the prison doctor rightly thought that endocrine treatment might help both their instability and their obvious physical infantilism. The examination of their hands proved to be something of a sensation. All the boys had the same pituitary infantile hand, true to type, which I have just described, and three of them possessed almost identical crease-line patterns, characterized by the simian line. There were only the three main creases present, practically no accessory lines, and all the creases were very deeply marked and 'badly' drawn. Only the fourth boy, who was of a slightly higher intelligence, had a more developed crease-line system. I demonstrated the hands to Dr. Simpson and both of us agreed that they showed what one may call a 'collective personality'. There were indeed very little individual differences between these boys. The clinical diagnosis was: hypo-pituitary infantilism.

The temperament which goes with this type of hand is characterized by instability of mood and narcissism; the latter leads to the desire to please, to show off and to be admired. These people are completely self-centred and with little or no regard for others. Homosexuality is frequent. I knew a number of these boys who were very graceful and handsome and a great success, especially with their superiors. The reader must not, however, get the impression that the owner of the infantile pituitary hand is only to be found inside 'institutions'. This is not the case. One meets the same kind of personality in ordinary life, but in a great many cases these people live on the fringe of the law.

3. THE PITUITARY-OBESITY HAND

There will be few among the readers of this book who have not, at one time or another, met a person who had the characteristics of the Froehlich syndrome. They are abnormally fat people, with bulging cheeks which almost hide their eyes. They always have a marked embonpoint, and small hands and feet. They walk very lightly in spite of their enormous volume, and carry their heavy bodies with unusual grace.

The Shape of the Pituitary-Obesity Hand

This is a pudgy hand of which the palm is much longer than the fingers. The palm is broad, rather because of the great amount of fat tissue than muscular development. The thenar- and hypothenar eminences are flabby. The fingers are most characteristic of endocrine disorder. They are always short, either tapered or pointed with sausage-shaped basal phalanxes. Noteworthy is the quality of the palmar skin, which is as soft as that of a baby.

The Nails. The nails are mostly very small or even underdeveloped. This is not always the case, however, and the nails cannot, therefore, be taken as a sure symptom of the endocrine condition. Even the small nails generally have moons. A characteristic feature is the shiny quality of the nails.

Physical Qualities. The outstanding feature is the very high flexibility of the fingers. The palm is in many cases flexible but I have also seen stiffness, even in hands with very flexible fingers. In all cases the terminal phalanxes can be bent backward, which is a trait of great importance for differential diagnosis. The thumb and often other fingers as well are usually double-jointed.

TEMPERAMENT AND MENTALITY

Placidity of temperament and *joie de vivre* make the owners of this hand delightful company. They are the favourites at school and in the family circle. They love to laugh and joke and never spoil a game. Their endearing nature is enhanced by lack of aggressiveness which does, however, put them at the mercy of other people. They cannot defend themselves well. They are very affectionate and helpful. They are, indeed, radiant children, also at a more advanced age. They frequently have a streak of obstinacy in them, however, but they always remain happy-go-lucky and children at heart. They are mostly of average intelligence; I have never found a person with the Froehlich syndrome among mental defectives. They love music and dancing, and there are fine singers among them. They are very light on their feet and are good dancers, in spite of their massive bodies. I knew a little girl with all the symptoms of pituitary obesity who was an excellent acrobat. All her fingers were double-jointed. Her mother, who showed the same endocrine make-up, was a first-class rider.

At the end of this description of pituitary hand-types I must point out that the hyperpituitary factor is comparatively rarely found in 'normal' people, while the hypopituitary factor, as represented, for example, in the different types of infantile hands, is present in a great number of people in all walks of life. The Froehlich syndrome in its pure form is also of rare occurrence, but the fat hypopituitary person is a well-known figure in every milieu and, I think, in almost every land. This goes to show that the endocrine factors, together with their emotional and mental concomitants, can be found in a great many people who have no clinical endocrine symptoms. But I also want to stress that conclusions on the mental make-up, drawn from clinical endocrine research, must not be projected wholesale on people with endocrine factors of the nature described. A factor is only a partial aspect, and other factors have to be taken into account to find the true psychosomatic formula of an individual who is not a 'fully fledged' endocrine case. There is, however, as I have already said, in every person and in every pair of hands, an individual endocrine factor which has to be weighed carefully with all the other factors determining the shape and other hand-traits in order to come to the right psychosomatic diagnosis.

The difference between normal and abnormal conditions is, however, only one of degree as I have pointed out in this chapter, as well as in my earlier writings. In order to illustrate how closely 'normal' and pathological endocrine functioning is connected, I want to relate the following example: I had diagnosed a tendency to hyperpituitarism (acromegaly) ten years ago in the case of a middle-aged man, and had sent him to an endocrinologist for the confirmation of my findings. The specialist ridiculed my diagnosis and found neither clinical symptoms (which was exact) nor was he convinced that the hands showed any marks of early acromegaly. They were in fact the very large hands of a man of over six feet, but I had observed that the phalanxes were beginning to thicken. The person concerned naturally believed in the verdict of the endocrinologist and nothing was done about it. He became more and more neurasthenic and exhausted and showed all the symptoms which are relevant to the mental and emotional attributes of hyperpituitarism. When he consulted another specialist seven years later my diagnosis was fully confirmed. It was then an easier job as the hands and feet had become very large and definitely oversized. A similar example happened to come my way in the case of a young girl developing myxoedema. I had diagnosed the condition from hand-features two years before clinical symptoms appeared.

B. THYROID HAND-TYPES

I. THE HAND OF THE CRETIN (CONGENITAL HYPOTHYROIDISM)

One of the well-known conditions of hypothyroid function is cretinism. The hands of the cretin are almost as familiar to the general public as the hands of mongolian imbeciles, which they resemble in certain aspects.

The Shape of the Hand

This is a broad and very short hand with stout fingers which have spatulate tips. The fifth finger is mostly undersized, pointed and crooked. The thumb is of an ugly shape, being too short with a plump and broad terminal phalanx.

The Nails. They are small, often under-developed and without moons. They are, in the examples at my disposal, striated, a quality which seems to be connected with hypothyroidism, as is the overgrown cuticula (the skin which grows from the nail-bed over the nail).

Physical Qualities. The tendency is towards stiffness of the palm and fingers. The palmar skin is very dry and the hands are habitually cold. The palmar skin is of a hard texture, which is a symptom interesting for differential diagnosis.

The Crease-lines. The creases are few in number and on a pattern which is found in mental defectives.

TEMPERAMENT AND MENTALITY

The temperament of the cretin is in some respects similar to that of the mongolian imbecile, but it differs in others. The cretin has the same responsive and affectionate disposition and is mostly the pet of his milieu. He is utterly defenceless because of his lack of aggressiveness and his inferior intelligence. He is slow in all his reactions and phlegmatic to the point of lethargy. Cretins are 'emotional' people, prone to anxiety and depression and have a great sense of their inferiority.

2. THE MYXOEDEMATIC HAND

The myxoedematic hand does not differ much from the hand of the cretin. The condition is only reflected in the shape of the hand, as described in the previous paragraph, when it occurs in early youth before the epiphyses of the bones of the hand are closed. I have observed one difference, however, namely that the hands are more pudgy than in cretinism and the basal phalanxes of the fingers are sausage shaped.

There is the same hard and dry palmar skin and rigidity of the joints of the hand and fingers, as well as similar characteristics of the nails and crease-lines.

According to Dr. Simpson, psychosis is frequently associated with myxoedema; this is an interesting observation which points to the intimate link that exists between endocrine function and the mind as well as the intelligence.

When myxoedema occurs in later life the hands are less, or not at all characteristic. In people who have been suffering from a slight form of hypothyroidism all their life, without symptoms of myxoedema, the hands are usually broad and short with sausage-shaped fingers. But that is by no means always the case as this endocrine dysfunction also occurs, for instance, in people who have been previously hyperthyroid.

3. THE HYPERTHYROID HAND

Hyperthyroidism rarely occurs in early youth but is a disturbance to which the middle-aged man and woman are subject. That is why one cannot exactly speak of a congenital hyperthyroid type of hand.

The shape of the hand in hyperthyroidism is conditioned by the age of the patient, the general state of his health and the endocrine condition before the onset of hyperthyroid function. As I have already said hyperthyroidism can be superimposed on a number of other endocrine dysfunctions, even those which go with symptoms generally found in connexion with endocrine under-functioning, as for example, obesity. If there is a marked deposit of fat in a hyperthyroid person, the hands are naturally rather broad with stout or even sausage-shaped fingers. If there is marked slimming in the hyperthyroid sufferer, the fingers are likely to be slim, with thin basal phalanxes. In a number of cases thyrotoxicosis appears in people of a leptosome, slim physique, a constitution which has, according to Sheldon, a tendency to hyperthyroidism. In these 'constitutional' cases the hands are characteristic.

The Shape of the Hand in 'Leptosome' Persons with Hyperthyroidism

The hands are long and bony with fingers with thin phalanxes; this characteristic is particularly marked in the basal phalanxes. The shape of the hands is similar to that which I have described in my book *The Human Hand* as the motor bony hand. The shape of the fingers closely resembles those of people who suffer from tuberculosis. In a study which I undertook on 125 tubercular patients (not published) I found this type of hand in 65 per cent of cases. It is a well-known fact that tuberculosis and hyperthyroidism are often found together, but one has to be guarded not to confuse the two conditions from similar hand-traits.

The Nails. The nails are long and often curved, of a shiny quality, with large and well-defined moons. This is probably due to the accelerated tempo of growth of the nail because of the accelerated metabolism of the hyperthyroid person.

TEMPERAMENT AND MENTALITY

The hyperthyroid person is nervous, apprehensive and anxious. His nervousness takes mostly the form of restlessness, often coupled with sleeplessness. There is generally an excessive amount of nervous energy available which is dispersed in too many different activities. Concentration is lacking. The people belonging to this group are generally of good or high intelligence, quick in the uptake and imaginative in their associations. But they spoil their great possibilities through their over-activity and their superficiality.

C. THE EUNOCHOID HAND

There are two variations of the eunochoid hand: the thin and the fat, as there are two variations of eunochoid types (see *Major*

Endocrine Disorders by S. L. Simpson). The eunochoid hand of the thin variety is an exaggerated form of the long sensitive hand which I have described in *The Human Hand*, and in *A Psychology of Gesture*.

The Shape of the Hand

Both palm and fingers are long, giving the hand an elegant air. Among women the thin variety is the rule, and among men the fat palm is found equally often, but both hand-types have the same long and tapered fingers.

The Nails. The nails are generally long or over-long; often curved and furnished with big moons.

Physical Qualities. Flexibility of the palm is, as a rule, normal, and that of the fingers considerable. The finger-tips can often but not always be bent backwards. Other physical qualities vary widely according to individual factors.

The Creases. They are equally numerous in vertical and horizontal direction. There are many accessory crease-lines which are superficially drawn.

TEMPERAMENT AND MENTALITY

The eunochoid person is biologically bisexual and in both men and women one finds features of temperament and mentality which are characteristic of the opposite sex. They are predominantly 'introvert' people of a nervous and sensitive disposition, with a weak emotional response and little sense of adventure because of their marked insecurity. They often have a well-developed aesthetic sense, love the theatre and ballet and show frequently an inclination towards mysticism. There is something frail and brittle in their make-up, which forces them into exaggerated self-protection.

On the whole, they are very intelligent people, versatile and gifted in the arts and crafts. They are often found among artists of an interpretative type, such as actors or they excel in fine crafts and fine mechanics.

D. THE ADRENAL HAND

Before speaking of an adrenal hand-type I have first of all to explain some aspects of adrenal functioning. The adrenal glands are small endocrine glands situated above the kidneys and consist of an outer cortex and an inner medulla. These names already indicate the intimate link which exists between the adrenals and the brain. Indeed, these glands are intimately connected with both the central and the autonomic nervous systems. One of the most important functions of the adrenal cortex is the secretion of adrenalin into the blood stream. In fact, the adrenals react in all emotional states and influence both emotional behaviour and expression.

Another outstanding feature of adrenal functioning is the stimulus to sexual function and development and the influence on nervous as well as muscular energy. When there is an over-activity of the adrenal cortex in prenatal life, the child has marked attributes of the other sex and pseudo-hermaphroditism results. Over-activity of the gland in early youth leads to sexual precocity as seen for example in the hercules boy. I had the opportunity to see the hands of both a pseudo-hermaphrodite and a sexually precocious child, a girl of the age of four. These are rare cases, but hyperadrenalism after puberty is of more frequent occurrence. In its pathological form it is called the adreno-genital syndrome which goes with virilism in women and with hypersexuality in men. The condition is, however, much more common in the female than in the male sex. One finds in hyperadrenal women not only the secondary sex characteristics of the male but also a great increase in muscular strength. As Simpson mentions in his much-quoted book, there are thousands of 'normal' women with a certain degree of virilism, which reveals a hyperadrenal disposition. In describing the adrenal hand I am speaking of a hand-type actually found only in women, who need not have any clinical symptoms of hyperadrenalism whatsoever. They are of a tall and athletic physique and of a masculine appearance. This physical trait is probably due to an over-activity of the pituitary adreno-cortical hormones; the two glands are functionally interrelated.

The Shape of the Hand

The hands are of a masculine type. They are broad and long. The bones are strong and the muscles, especially those which compose the thenar- and hypothenar eminences, are extremely well developed. The fingers are long and strong with either square or spatulate finger-tips.

The Nails. The nails are well developed; generally of medium size and rather flat with small moons.

Physical Qualities. These are predominantly stiff hands. They are generally moist and cold, but of normal colour.

The Creases. They are broad and deep and accessory crease-lines appear in comparatively great quantity.

TEMPERAMENT AND MENTALITY

The hyperadrenal factor always produces great irritability, nervousness and a lack of emotional balance. One may speak of emotionalism as characteristic of this temperament. Emotional disturbances and neurotic conflicts are to be expected in this type of personality which is, by disposition, bisexual, but rather strangely not frequently homosexual. One finds, on the contrary, that hyperadrenal women are sexually orientated towards men. They are dominant and authoritative people and very often domineering. The mentality shows masculine attributes and interests. Manual skill and artisanship is frequently found and these women are excellent sportswomen. They are good drivers, good equestrians and often first-class shots. They also excel in games and athletics. Their intellectual level is, as a rule, not particularly high. They have organizing gifts as well as good business sense and make perfect manageresses. They don't like, on the whole, petty talk and gossip and are loyal and reliable friends in spite of their emotionalism.

The description of the adrenal hand with which I conclude this chapter, shows how very closely pathological endocrine conditions are linked with endocrine dispositions which are normal. The same holds good, as I have already pointed out, for other endocrine tendencies, the pituitary infantile hand being a good example of the frequency of a tendency towards hypopituitary functioning in 'normal' people. But the same holds good, in fact, for every endocrine hand-type mentioned in this chapter, as the difference between normal and abnormal endocrine functioning is only one of degree and intensity.

CHAPTER III

THE HAND AND TEMPERAMENT

IT is common knowledge that to test the temperament is a very difficult problem in psychology, for emotions are irrational and escape direct measurement. Subconscious urges determine the emotions and through them—behaviour. Two underlying factors can be differentiated in both emotions and behaviour: the constitutional and the environmental. The two are of course interrelated, and it is sometimes difficult or impossible to separate them from each other. It must be made quite clear that constitutional tendencies are not entirely 'congenital', but are up to a point conditioned by the psycho-physical evolution of the child. This evolutionary development in the first years of life shows the most complete form of interaction between the ME and the world. The child lives in much stronger unity with his surroundings than the grown-up person. Certain purely physiological happenings may also be decisive in shaping temperament and behaviour. It suffices to mention the influence which certain infectious diseases can have on the endocrine glands, and through them on temperament. For example, one form of hypopituitary infantilism, called after its discoverer the Levy Lorraine Syndrome, is produced through early tubercular infection affecting the pituitary gland as well as the whole hypothalamus region and resulting in a change of emotional reactions which alter the psycho-physical constitution for ever. And, indeed, any other illness or shock which involves this region of the brain inevitably leads to a specified reaction- and behaviour-pattern which becomes constitutional. This is definitely so when the injury occurs in early childhood, but even in puberty and after, existing constitutional trends can be changed by any injury affecting the pituitary-hypothalamus region.

My subject deals with the constitutional factors of temperament as they are the only ones which can possibly, in one way or another, be seen in hand-traits. Before I begin to show if and which correlations exist between the characteristics of the hand and the characteristics of the emotional make-up, it is necessary to make some preliminary observations.

1. *THE IMPORTANCE OF MAKING OBJECTIVE TESTS OF TEMPERAMENT*

I have already stressed in my book *The Human Hand* how important it is to have a test of temperament not only to acquire a more profound and greater knowledge of Man but for psycho-therapeutic reasons. If any method could be found for discovering the emotional make-up other than the subjective report on himself by the person concerned, the psycho-analyst would possess most valuable insight into the background of neurotic disturbances. This view is strongly supported by Sheldon in his article 'Constitutional Factors in Personality'. It is hardly necessary to mention other advantages of an objective test of temperament, for example in the selection of children for special education or of students and employees for professional purposes, etc.

I think no one can doubt the tremendous importance of the use of a test of temperament which is both objective and quick, but it must be made clear from the start that it is impossible to use any single method as a master key to lay open the mysteries of the whole personality. In order to get an all-round picture of all the factors which compose the temperament, and of those which determine personality, several tests must be made.

2. DEFINITION OF INSTINCT AND EMOTION

Before attempting to go into the heart of the matter and to show how hand-features are correlated with the emotional make-up, or the temperament, I must first give a definition of emotion as distinct from instinct.

(*a*) INSTINCT

Instinct is a primitive urge on an automatic level, which releases immediate action. Through this action complete or final satisfaction of the instinctive urge is reached.

(*b*) EMOTION

Emotion is a state of mind which goes with physical sensations that tend to increase with the strength and duration of the emotion. Emotion, especially in its cruder forms, has always an element of conflict and pain. Even the pleasurable emotions contain a feeling of nostalgia. William James differentiates the coarser from the finer emotions, the latter being akin to feeling, as for example aesthetic appreciation or love of nature, while the former are akin to instinct.

Emotions express themselves first and foremost in expressive behaviour, namely gestures and postures. The psycho-motor apparatus is the field of emotional expression. The sub-cortical parts of the brain are activated in states of emotion and especially the hypothalamus region reacts under emotional stimuli. This part of the brain is closely connected with the functioning of the autonomic nervous system which responds to any emotional fluctuation. The endocrine glands, and particularly the adrenals are also stimulated by emotion.

A more complete definition of emotion can be found in my book *A Psychology of Gesture.*

3. SOME DATA ON THE DEVELOPMENT OF CONSTITUTIONAL PSYCHOLOGY

I must also give some data on the development of constitutional psychology, to which my own subject belongs.

The correlation between the physical constitution and temperament has been recognized since the days of Hippocrates, who also pointed out that different psychological or 'temperamental' types have tendencies to different types of illnesses, an idea to which a whole branch of science, psychosomatic medicine, is devoted in our day.

Constitutional psychology developed rapidly after the First World War through outstanding contributions to the subject made by such protagonists as E. Kretschmer and W. H. Sheldon. Kretschmer owes a lot in his method of investigation to the Italian psychologists Viola, Naccarati and de Giovanni. I spoke at length of Kretschmer's contribution to constitutional psychology in my book *The Human Hand*, in which I still followed his typology in the interpretation of my hand-types. Here I shall sketch only briefly the main results of Kretschmer's work. Kretschmer differentiated four morphological types which he found correlated with two types of temperament :

MORPHOLOGICAL TYPES	TYPES OF TEMPERAMENT
(*a*) *pyknic*	*cyclothyme*
(*b*) *athletic* }	
(*c*) *leptosome* }	*schizothyme*
(*d*) *dysplastic* }	

The typology of Kretschmer is built on mental cases. He observed that patients with manic-depressive psychosis are of a pyknic physique, while schizophrenia goes in the majority of cases with leptosome-asthenic, athletic and dysplastic physiques. I am not discussing here the value and shortcomings of Kretschmer's typology; but I must mention that I have already digressed from his findings in my book *A Psychology of Gesture*, and my conception of psycho-physical constitutional types has in the meantime gone even farther away from Kretschmer's ideas.

Apart from Kretschmer, W. H. Sheldon is undoubtedly the most important worker in the field of constitutional psychology. He is the much more modern research worker of the two, which is probably due not only to a difference in age but also to the fact that he is an American and Kretschmer a German. The importance of the 'isolating' factor in German psychological research cannot be overestimated; the opposite, namely complete freedom both materially and mentally, has given American authors an extraordinary chance to advance, and they are in many respects far ahead of research workers in other countries. This does not mean that I agree with the often too mathematical, too mechanical approach of American psychological methods. I do not think that this specific danger applies to W. H. Sheldon. He is, undoubtedly, the most important worker on anthropometric studies of personality. He does not believe in psycho-physical types but assumes that there are a number of morphological factors which appear in a great variety of combinations and which determine both human physique and personality. He differentiates three fundamental factors:

(i) *the endomorphic*
(ii) *the mesomorphic*
(iii) *the ectomorphic*

He has devised a scale of seven points for every factor to express their intensity and importance in the ensemble of the personality. The theory and the method of what is called somatotyping is found in *Varieties of Human Physique* by Sheldon, Stevens and Tucker. I am giving here only a short description and definition of the three morphological factors from Sheldon's article 'Constitutional Factors in Personality' which appeared in *Personality and Behaviour Disorders*, Volume I (1944):

When endomorphy predominates, the digestive viscera are massive and highly developed, while the somatic structures are relatively weak and undeveloped. Endomorphs are of low specific gravity. They float high in water. Nutrition may, of course, vary to some degree independently of primary components. Endomorphs are usually fat but they are sometimes emaciated. In the latter case they do not change into mesomorphs or ectomorphs any more than a starved spaniel will change into a mastiff or a collie. They become simply emaciated endomorphs.

When mesomorphy predominates, the somatic structures (bone, muscle and connective tissue) are in the ascendancy. The mesomorphic physique is high in specific gravity and is hard, firm, upright and relatively strong and tough. Blood vessels are large, especially the arteries. The skin is relatively thick with large pores and it is heavily reinforced with underlying connective tissue. The hallmark of mesomorphy is uprightness and sturdiness of structure, as the hallmark of endomorphy is softness and sphericity.

Ectomorphy means fragility, linearity, flatness of the chest and delicacy throughout the body. There is relatively slight development of both the visceral and somatic structures. The ectomorph has long, slender, poorly muscled extremities with delicate pipestem bones and he has relatively to his mass the greatest surface area and hence the greatest sensory exposure to the outside world. He is thus in one sense overly exposed and naked to his world. His nervous system and sensory tissue have relatively poor protection. It might be said that the ectomorph is biologically extraverted as the endomorph is biologically 'introverted'. Psychologically, as we shall see later, these characteristics are usually reversed, the ectomorph is the introvert, the endomorph is one of the extraverts. The hallmark of ectomorphy is the stooped posture and hesitant restraint of movement.

Sheldon's idea of variable factors constituting human physique is based on a physiological conception, namely that of embryonic tissue. The organism is built up on three different 'membranes': the endoderm, the mesoderm and the ectoderm which, in the process of embryonic development, are transformed into the different sets of organs constituting the body. It is a very sound and convincing idea to relate the fundamental factors of constitution right back to their origin. This conception is the nucleus of the whole work of Sheldon. The method which results from this ingenious idea, and its application, are a natural consequence. The very flexible scale of mixtures of the three fundamental factors is an innovation in constitutional psychology which represents a turning-point in its development. In *Varieties of Human Physique*, Sheldon has lifted the basis of constitutional psychology out of a static and more or less 'unreal' atmosphere into dynamic flexibility and reality. In this respect the importance of Sheldon's work cannot be too highly stressed, while some criticism must be made of certain psychological terms which he uses.

In his later work *Varieties of Temperament*, Sheldon and his collaborator Stevens have devised a scale of temperaments correlated with the three morphological factors. Human temperament is just like human physique the result of a mixture of different factors which appear in individual intensity and variation in every human being. Sheldon differentiates a

(i) *viscerotonic*
(ii) *somatotonic and*
(iii) *cerebrotonic*

temperament, and has listed twenty traits which are characteristic of each temperament, in three columns. These traits vary from psychomotor characteristics to desires for congenial living and characteristic occupations. While it is easy to imagine a person with traits defined in these lists, I cannot help finding them muddled, or at least, ill-defined. Traits are listed which have nothing to do with temperament whatsoever. Nor is any attempt made to give a definition of temperament, the conception of which is purely descriptive and too much open to chance. Sheldon derived the characteristics of each temperament from psycho-analytical experience. He himself analysed a comparatively small number of students of different somatotypes and made notes of their emotional and other reactions.

Sheldon uses the terms extravert and introvert in his scales of temperament, and here he is also, in my opinion, much less clear and independent than one would expect. It was C. G. Jung who differentiated the extravert- and introvert-personality types in his book *Psychologische Typenlehre*. Kretschmer's classification of cyclothyme and schizothyme temperaments corresponds closely to Jung's types.

I have come to the conclusion, both from theoretical considerations and from practical experience, that the division of human beings into extraverts and introverts does not hold water, even when it is admitted that there are many transition cases, as Jung naturally does. Extraversion and introversion cannot designate a personality, as both tendencies are indicative of 'movement'. One 'inner movement' is going intrinsically towards the impressions of the outside world (in the extravert personality) and the other 'inward movement' is going intrinsically towards the imagination or the inner world (in the introvert personality). Extraversion and introversion are, therefore, tendencies which, on the whole, go with a certain structural make-up but which are, because of their very nature, flexible attributes and are more open to changes than other attributes of a personality. Extraversion or introversion can be markedly changed by environmental influences in both the child and the adult. Introversion can be produced in a predominantly extravert child, for example, through educational mismanagement or simply through loneliness.

I hope to have made it clear that it is possible only to speak of extravert and introvert tendencies and of those only with reserve and without finality. It seems to me quite clear that, as a rule, every person possesses both extravert and introvert urges, the intensity of which varies individually. Complete introverts are only to be found among mental patients and imbeciles, where any form of reaction towards the milieu is gone or completely deformed.

4. DEFINITION OF TEMPERAMENT

Temperament is a composite entity in which the following factors can be discerned:

(*a*) *the endocrine factor*
(*b*) *the psycho-motor dispositions*
(*c*) *the factor of sensitivity*
(*d*) *the factor of inhibition*
(*e*) *the factor of imagination*

There are certainly other factors not mentioned in this scale which are of importance in the emotional make-up, but these five appear to be sufficiently tangible to lead to some grasp of the structure of temperament itself.

5. DISCUSSION OF THE DIFFERENT FACTORS OF TEMPERAMENT AND THEIR CORRELATIONS WITH HAND-TRAITS

(*a*) *The Endocrine Factor*

This has been dealt with in a chapter apart and needs no further mention.

(*b*) *The Psycho-Motor Dispositions*

These can be divided into postures and gestures. There can be no question that both reveal emotions as well as the emotional make-up.

I am of the opinion that psycho-motor dispositions are intrinsically linked with both temperament and intelligence. I have dealt with the correlation of psycho-motor habits and intelligence in the first chapter of this book. To understand fully the link between psycho-motor behaviour and temperament it is necessary to go back to the evolution of brain functions in the child. As H. Wallon explains in his book *De l'Acte à la Pensée*, the child goes through an apprenticeship in regard to all his psycho-motor manifestations, which is due to the organic maturation of the brain. In doing so, the child advances at the same time in his emotional and mental development, or, in Wallon's own words: 'Ainsi se poursuit la croissance de l'enfant, non pas selon ce qu'il est a chaque instant mais selon le type qu'il doit realiser comme adulte.' Wallon thus points out that there is a core, or 'gestalt', of personality in the child which contains all his potentialities, mental as well as emotional. Individual psycho-motor reactions belong to the personality as closely as the eyes, the nose and the hands belong to a human body. Psycho-motor reactions which manifest themselves in postures and gestures are the vehicle of the emotions and can be looked at as a 'script' from which the temperament can be read in its individual variations.

Obviously, temperament can be seen in psycho-motor behaviour, but can it also be seen in hand-features? I have worked out in my book *A Psychology of Gesture* five fundamental psycho-motor patterns from observations made on expressive behaviour of mental patients. I have shown at the same time that these patterns are correlated with hand-types. My diagnosis of temperament and personality from both gestures and hand-traits corresponded very closely to the diagnosis made by the psychiatrists of the mental hospital concerned. From this it can be concluded that the temperament can be seen in hand-traits as well as in gestures and postures.

Another question arises, namely, what is the advantage of diagnosing temperament from the hand when one can diagnose it from gestures and postures? In order to get an idea of the temperament of a person via his expressive behaviour much observation is necessary, just as a painter needs a great number of sittings to paint a portrait. Expressive behaviour reveals first and foremost the *actual* emotions and an immediate emotional situation. The fundamental emotional disposition, or the temperament, can only be indirectly deduced after careful scrutiny. While a diagnosis from expressive behaviour must necessarily always emphasize the *actual* psycho-dynamic situation, the hand shows nothing of this but reveals the *constitutional emotional disposition* for the following reasons:

In the first chapter of this book, 'The Hand and Intelligence', I explained that the crease-lines of the hand are very largely the result of psycho-motor habits which manifest themselves from the earliest days of life, perhaps even in prenatal days. Thus the crease-lines represent the imprint, so to speak, of the sum total of all hand-movements, from the most simple to the most subtle ones, which are, to start with, unconsciously performed and later on, together with the development of the intelligence, obey both cortical and subcortical impulses. The way these movements or gestures are executed, their rhythm, their speed, their co-ordination, their range and individual variety depend on:

1. *The innate emotional disposition*
2. *The intelligence*
3. *The strength of the nervous dynamism*

Although intelligence controls more and more hand-movements in the course of mental development, all expressive movements are of an emotional nature and are subconsciously performed. The emotional factor remains decisive in psycho-motor expressions through the whole of life, even after the intelligence has taken the lead in many ways in the constant fight of man to adapt himself to his environment. Undoubtedly, the intelligence has a marked influence on psycho-motor behaviour, but it remains, even in the adult, only a regulating power. The real motivations of psycho-motor expressions are always, even in the 'detached' personality, emotional in their source.

Psycho-motor habits are mostly expressed by hand movements and postures and this in itself is evidence that a close link exists between the emotional make-up and the hand. While the structure of the latter, by which is meant the development of the bones, is determined by endocrine factors, the formation of the creases is mainly the result of earliest psycho-motor habits expressed by hand movements. The fat tissue and certain qualities of the skin are related to endocrine function, particularly to the working of the thyroid gland. The muscular formation of the palm as well as the quantity and quality of the crease-lines are related to movements and are characteristic of psycho-motor habits for the following reasons: the hand is an instrument of two types of movement: (i) movements of purpose (cortical); (ii) movements of expression (subcortical). If the former prevail the palm is broad and muscular with few and deep crease-lines, predominantly in a horizontal direction. If the latter prevail the palm is flat with superficial crease-lines in both a horizontal and vertical direction. The broad and fleshy palm, which generally goes with short and strong fingers, belongs to a personality with predominantly extravert tendencies, while the flat palm with long and thin fingers is characteristic of the personality with predominantly introvert tendencies. As Gurwich (1939) has pointed out the prevalence of movements of flexion (cortical) designates well-centralized personalities, while the prevalence of movements of extension (subcortical) are typical of badly integrated personalities.

It seems evident that different personality 'types' go with different temperaments which can be diagnosed from the shape of the hand and its muscular pattern.

The formation of the muscles of the hand is in the first place constitutionally conditioned, but their actual development is due to the constant exercise of the hand in movements and to its muscular tension and relaxation in posture. There is no moment in anybody's life when the hand is not 'active' either in gesture or posture. To gestures also belong the hardly visible, very subtle, modulating movements of the fingers which are performed practically all the time even during sleep.

It is thus obvious that the character of psycho-motor patterns, which are part and parcel of the temperament of a person, can be divulged from the muscles and crease-line pattern of the hand. As both muscles and creases are essentially shaped in the first years of life, which is the age of the emotional development, they can be considered *the crystallized imprint of temperament.*

(c) *The Factor of Sensitivity*

By the factor of sensitivity I understand the degree of receptivity to outside impressions. Sheldon attributes the highest degree of sensitivity to the man with the strongest ectomorph factor, whose temperament he describes as cerebrotonic. This classification corresponds in essence to a person with a long sensitive type of hand which goes with the highest degree of ectomorphy. But why and how does the factor of sensitivity show in hand-traits? To explain this I have again to go back to the evolutionary history of the child. Sensitivity depends on perception. Any form of perception leads to one form or another of psycho-motor reactions. The perception through the sense of touch, as localized in the hand, is always and automatically linked with movement. One speaks of perceptivo-motor reactions to signify that perception and movement are an inseparable unit. The hand is the seat of different forms of sensibility, as for example for hot and cold, pressure, pain, light sensations of touch and a deep muscular sensibility. Sensitivity to light touch is most developed in the finger-tips. It is this concentrated sensibility of the fingers which is both the source and the channel of all sensitivity. Collins and Drever point out in their book *Experimental Psychology* that cutaneous sensibility represents the most primitive of all the sense departments, and that all the other senses may be looked upon as developed from a primitive sense of touch. As mental images have their primary root in the knowledge of objects of the outside world, so even the most subtle human sensitivities have their root in sensations of touch. The child makes itself familiar with its own body and the outside world in the first place via his sense of touch localized in the hands. The sensibility of touch gives rise to motor reactions, which in turn are responsible for certain essential qualities in the mental representation of images. The spatial sense, which is also most concentrated in the hand, is one of the most important factors in the development of the intelligence and the imagination, as I have already pointed out in the first chapter of this book. The representation of objects in space leads to the conception of objects as tridimensional. The spatial sense gives distinction and perspective to mental imagery, and apart from that is responsible for the sense of reality, both physically and mentally. It is the foundation of physical and mental balance. It can thus be said that sensitivity towards the world and towards the Self has been built up by means of the hands. The child learns through his sense of touch and through imitation. The power of the latter seems to be unlimited in the child. It can animate dead objects: a stick may represent a horse, a doll represent a child, a chair a mountain, or a number of other things. The child imitates also the emotions of his milieu, first of all automatically, then sympathetically until he has incorporated them into his psycho-physical organism and they have become his own. From then on he expresses emotions more than he imitates them, and lives from a new centre—his own personality. The chain of development into an individual emotion reaction-type starts with the different forms of sensibility, particularly with the tactile sense of the hands, and leads through psycho-motor imitation to the creation of the individual temperament.

As the hands are the tools of 'feeling' as well as of emotional expression, it can be assumed that the capacity of feeling, or the factor of sensitivity, can be seen in hand-traits. As the fingers, particularly the finger-tips, are the instruments of primary sensibility, it is in their shape and other characteristics that the factor of sensitivity can be traced. In people in whom sensitivity is highly developed, one can observe a wide range of subtle and delicate movements of the fingers. These gestures are easily stimulated by outside impressions as well as by the imagination. The potential range and variety of 'sensitive' gestures cannot only be deduced from characteristics of the fingers, but also from certain crease-lines, namely those which are the result of subtle finger movements, particularly the creases on the distal part of the palm. The left hand is more significant for 'sensitivity' than the right, as I have been able to show in my book *A Psychology of Gesture*. Expressive reactions tend to be performed by the left hand in the first place, while the right is mainly used for purposeful movements.

There are also other reasons for assuming that both the degree and quality of sensitivity are much more marked in the left than in the right hand. In my book *The Human Hand* I have given some proof of this which corresponds closely to the views, more recently expressed, in the French psychological journal *L'Enfance* by Madame Roudinesco and Jean Thyss (in *L'Enfant Gaucher*, 1948). The authors point out that in all organs which exist '*en double*' there is a division of function, which means in the case of the hands that the right takes over the active, and the left the passive, or feeling side.

En effet, la main la moins habile du point de vue moteur a de ce fait même un role sensitif particulier de palpation des objets tenus par la main droite ou d'exploration de l'espace ouvrant la voie aux activités motrices de l'autre main. Dans ce cas-là la main droite sentira par l'intermédiare de l'outil, tandisque la gauche tatera et saisira directement. Il faut se souvenir de cette intelligence tactile et stéréognostique de la main mineure en étudiant les apraxies et les alterations du sens de l'espace par atteinte du schème corporel.

I have been able to demonstrate (see *The Human Hand*) that there is normally a difference in the size of palm and fingers in the right and left hand, a fact that also points to their different functions and significance. The capacities of man which go with conscious control are linked predominantly with the right hand and those which belong to the domain of the subconscious are linked more with the left.

Apart from the characteristics of the hand I have just mentioned, the physical qualities, such as the smoothness or dryness of the skin, its habitual warmth or coldness, moisture and temperature have also a psychological significance for sensitivity. As these qualities are connected with the working of the automatic nervous system, they are, however, mainly linked with the immediate emotional state of the person concerned.

(*d*) *The Factor of Inhibition*

There is no mental or emotional activity without the factor of inhibition. Indeed, without inhibition there would be no inner life. In its repercussions on the whole psycho-physical organism of man the tension-relaxation theory of the emotions, so clearly explained by K. Goldstein, entails a fundamental truth about life processes altogether. N. Vaschide pointed out in his book *Essai sur une Psychologie de la Main*, that all life processes are the outcome of a conflict between a forward impulse and withdrawal. The factor of inhibition is implicit in what he calls withdrawal, and is one of the most important pillars of the structure and economy of our inner life. We are investigating here the part inhibition plays in the emotional make-up and how this factor, if at all, shows in hand-traits. When one speaks of somebody as being inhibited, one means that his emotional and mental economy is impaired through too much inhibition. I must make it clear that there is no person alive with a complete lack of inhibition which would, indeed, neither be possible nor desirable. Inhibition can be defined as a halting emotional tension. There is a functional link between emotional and nervous tension on the one hand, and postural behaviour on the other. The latter obeys the finest modulations of muscular tone. In order to make this link understandable I am quoting a passage from H. Wallon's book *De l'Acte à la Pensée*, p. 153 in French, and in my English translation:

> L'excitation ne reste pas seulement cérébrale. Elle se répand dans les muscles. Bien qu'ils puissent n'exécuter aucun mouvement, ils sont neanmoins le siège d'une activité parfois intensément ressenti par le sujet, bien qu'imperceptible pour les autres. Cette impression n'est pas une pure illusion. Ce qui donne lieu a pu etre constaté, enrégistré. C'est une contraction des muscles qui n'entraine pas leur raccourcissement, mais qui les tient au contraire dans seur forme présente et leur donne un dégré variable de consistance, de résistance aux déplacements. C'est leur fonction tonique.

The stimulation does not remain only cerebral. It extends into the muscles. Even when the latter cannot react by movement, they are the seat of an activity, which is sometimes most intensely felt by the person concerned, even if this activity is imperceptible to others. This impression is by no means an illusion. One has been able to check and register what is happening. It is a contraction of the muscles which does not lead to their shortening, but which keeps them on the contrary in their actual form and gives them a variable degree of consistency and of resistance to movement. This is their tonic function.

Wallon has thus described some form of muscular stimulus which is not discharged in movement, and which therefore represents an inhibitive factor. Inhibition increases muscular tone and produces muscular tension. If this happens to a degree that the natural tension-relaxation balance of the organism is disturbed, then the inhibitive factor is abnormally high. But as everybody knows, inhibitions affect not only postural behaviour but also gesture. The natural flow of gestures and other motor activities becomes disturbed. Movements become abrupt, halted and stilted. Accumulated muscular tension, produced by inhibition, can lead to emotional as well as muscular stiffness, and the latter may be taken for fibrositis. Inhibitions are either automatic or produced by thought processes. Both types of inhibition belong to the emotional sphere and therefore to the temperament of a person. The degree and intensity of individual differences in emotional inhibitions can best be studied in small children of similar or identical surroundings. The factor of a strong inhibitive mechanism narrows the field of emotions and limits human contacts; this can already be seen in the child. Strongly inhibited people isolate themselves and end up as eccentric bachelors or old maids. Kretschmer made the factor of strong inhibitions the principal feature of his 'schizothyme' temperament, and Jung does the same in the description of his introvert personality. Sheldon finds marked inhibition in what he describes as the cerebrotonic temperament. It appears from all investigations on the subject that the strongly inhibited person is at a great disadvantage and that the little inhibited person has, so to speak, the world at his feet. If one reads, for example, the three columns of Sheldon's description of temperaments, or if one peruses what Kretschmer and Jung say about the inhibitive factors one cannot help feeling that it amounts almost to a tragedy not to possess a cyclothyme temperament, or not to possess extravert tendencies only. According to Sheldon, most people would, if they could, choose a somatotonic temperament, as that seems to entail all that one can possibly wish for. I have for a long time mistrusted this obvious want of balance and I think that there is a subconscious bias in the evaluation of temperament by most psychological investigators. It may be that the person with predominantly extravert trends and a low inhibitive factor has a happy disposition, but that does not mean that the opposite type of man is necessarily of an unhappy nature. It only means, according to my view, that there are different forms of self-fulfilment, which call for different forms of living. An eccentric bachelor or a lonely old maid may have a great richness of enjoyment in their more impersonal or sublimated pursuits than the happy-go-lucky fellow. The person with little restraint tends towards hysterical reactions which spoil his or her serenity and impede any maturity in outlook and feeling. Happiness cannot be measured on a general scale. Anybody, to whatever group of temperament he or she may belong, has his or her individual chance of happiness within the limits of his or her make-up. Serenity, which is the ultimate goal of self-fulfilment, and happiness depend on balance and wisdom, of which the part wisdom plays actually comes first. And both are in the orbit of any temperamental make-up.

It cannot be denied that people with an abnormally high inhibitive factor are more prone to neurotic disturbances than the opposite type of person, and for this reason psychologists deal more often with the strongly inhibited personality. While the badly balanced individual with marked extravert tendencies has a disposition to hysteria, his opposite number inclines to suffer from neuroses of a psychasthenic type, such as obsessional- and anxiety-states. It is because of a higher frequency of nervous disturbances, which leads the 'introvert' into the analyst's consulting-room, that the bias about his 'unhappy' make-up has arisen in the mind of the psychological research worker.

The hysterical person sees himself as a 'wonderful' man or woman who always knows better than anyone else and to whom nobody can teach anything. He or she is self-righteous and obviously not inclined to seek easily a psychologist's help, as he is lacking in insight. Not so the psychasthenic person. He constantly seeks support from others. He is anything but self-sufficient and is fundamentally insecure, which makes him either inactive or impulsive or both. He often develops paranoid or obsessional traits. The ill-balanced 'extravert' thinks that he is coping beautifully with his life even when he brings himself and others into the greatest muddle. He remains over-confident in himself, while the ill-balanced 'introvert' easily loses heart and exaggerates his own shortcomings. This is particularly so in people with a fine intelligence and a natural analytical tendency, which is always strong in the person with predominantly introvert tendencies. Because of his insecurity, diffidence and anxiety, the man with an abnormally high inhibitive factor needs help most, and for this reason alone it is of very great importance to know about the strength of the inhibitive factor in his make-up in order to estimate what powers the person concerned may have to compensate for it.

How can the inhibitive factor be divulged from hand-features? To answer this question I must first go back again to psycho-motor behaviour. The inhibitive factor shows itself in tight postures of the hand and in gestures either of a staccato type or of over-emphasis. The too-short 'tired' gestures and abrupt movements of the inhibited and anxious person are well known to anybody who observes human expression. Movements of inhibition are lacking in rhythm and co-ordination. As the emotional qualities which are at the basis of these psycho-motor manifestations already exist in early childhood, it is evident that they are reflected in hand-traits.

The hand thus tends to mirror the constitutional factor of inhibition in everybody because of the natural process of evolution, but there is also a connexion between endocrine functioning and inhibition. The 'eunochoid' person, to take an extreme example, has always a high degree of sensitivity and a high inhibitive factor. Both must be innate qualities as eunochoidism is already prenatally conditioned. The long and thin hand of the eunochoid is an exaggerated example of what I have described as the long sensitive type of hand in earlier writings. The owner of such a hand has a great amount of the cerebrotonic temperament in him which is marked, as Sheldon has pointed out, by strong inhibitions. One can therefore say that the long sensitive hand is the hand characteristic of inhibition. The next to follow in this respect is the bony motor type of hand. It is on a similar pattern as the long sensitive hand but of a stronger structure and with better-developed muscles. It is found in people with a lean and slender physique who are good at games and at certain sports like golf or tennis. They are, as a rule, not good athletes. They are very timid and often eccentric, make little contact with their surroundings and are inclined to be escapists. The structural build of the hand is, as I have already said, determined by endocrine functioning and it is probable that both the long sensitive and the bony motor hand reveal a certain amount of subgonadism. The crease-line patterns on the palms are due to psycho-motor habits and one finds the flat palm connected with a tendency towards movements which express unease and withdrawal. I have observed that in the leptosome or bony motor hand the distal part of the palm is always flat even when the thenar- and hypothenar eminences are well developed, which may indicate a restricted pattern of subtle finger-movements in the person concerned. This means psycho-thinking.

Apart from general characteristics, such as the shape and the muscular development of the hand, the factor of inhibition can be seen also in the crease-line system, and, up to a point, in the physical qualities of the hand.

The crease-line pattern of any human palm is entirely individual and even that of the right hand and the left of the same person are never alike. This extreme individualism must mean something. There are many more secondary creases in hands which go with a strong inhibitive factor than in those which belong to a person with predominantly extravert tendencies. One of the principal characteristics of the sensitive types of hand is the great number of crease-lines and their superficial design. There are also more broken creases and other faults of design in these hands than in other hand-types, and both these characteristics are significant for a strong inhibitive factor. To a somewhat minor degree the same applies to the bony motor hand in which the accessory creases are very numerous but deeper than in the sensitive types of hand. The bony motor hand belongs to people with a marked ectomorph and mesomorph factor; in them the inhibitive factor is compensated for by a strong physical and nervous dynamism. It is this inhibitive factor, apart from a general nervous disposition, which is responsible for the quantity and quality of the creases, especially for faults in design. If inhibitive psycho-motor reactions are at work from early youth, the sum-total of psycho-motor manifestations is bound to show a zig-zag curve instead of a more or less straightforward line. All movements which don't come off properly, which are either halted or over-emphasized or badly co-ordinated, must in the long run reflect themselves in a faulty design of the crease-line pattern. The reader must not, of course, imagine that this is a process which could be watched from day to day or from year to year in the hands of children. The creases are, on the whole, established in their general design at a fairly early age, which, in my opinion, shows how early the structure of temperament and personality takes its definite shape. As Valentine said in his recent paper 'Trends, Dangers and Possibilities in Psychology' (*British Journal of Educ. Psychol.*, 1948), 'I also found highly specific traits even before the age of two which remained characteristic of the individual when grown-up.' There are, however, changes in quantity and quality in accessory creases, and even in principal creases until after puberty and often later. These are very small changes as a rule but they prove that a change of emotional reactions influences the habitual gestures and postures of a person to such an extent that new accessory creases appear on the palm, or very superficial ones may disappear.

The psychological significance of the physical qualities in general and for the factor of inhibition in particular is a well-known fact. The pale hand of the melancholic is familiar not only to the psychiatrist but also to the layman. The red-blue hand of the schizophrenic is also common knowledge, as are the moist and cold hands of the shy and inhibited adolescent. Coldness of the hand goes with depression or unease if there are no purely physical reasons for the symptoms described. Colour, temperature and humidity all express the inhibitive factor. And so does the degree of flexibility of the palm and fingers. In all cases where there is no other physiological cause, such as rheumatism or any other illness causing stiffness of the hand, bad flexibility is a sign of a strong inhibitive factor, as I shall show statistically in the clinical studies of this book.

As the physical qualities of the hand largely depend on the working of the autonomic nervous system, which is the direct transmitter of emotional reactions, it is easy to understand that they represent a kind of barometer of the inhibitive factor.

(*e*) *The Imaginative Factor*

Emotion and imagination cannot be separated from each other. There is no emotion which is not tinged and sustained by the imagination. And there is no imagery which is not in one way or another stimulated and kept alive by emotion. The imaginative factor can thus be considered as part of the emotional and the mental world. Naturally it is not possible to speak of a person's imagination before a certain age, before there is any coherence of mental processes. This happens at about the age of 4. From that time onwards till the age of 7 the child lives mostly in his imagination and in his emotions. This period of life goes with the first creative outburst which expresses itself in games, in drawing and painting, &c. One can observe that already at that age the imagination acts as an outlet for the emotions and develops at the same time the personality, which becomes more and more independent of the milieu the more the child lives in a world of his own. This process builds up his powers and his self-respect. He becomes the master of his small inner world. William James speaks of an imaginative Ego which compensates for the failures encountered in the real world. It is a well-known fact that only children and those who are frustrated or mismanaged by their elders develop their imagination very early and very readily and use it to counteract the shortcomings of their lives. H. Wallon points out in his book *L'Enfant Turbulent* that the imagination is a substitute for psycho-motor actions and can, indeed, replace action altogether. With this statement Wallon puts the imagination in line with the inhibitive factor, as the imagination inhibits motor action. The former represents a safety valve against the sufferings which any form of action, psycho-motor or other, could entail. The development of the imagination ensures the development of self-protection, and gives us the chance of escape at any moment and in any circumstances. But apart from this, the creative power of the imagination is the very groundwork and backbone of any individual richness as well as of the richness of any cultural pursuit. I think that the imagination more than the intelligence is the essence of human progress in every field. The evidence that the imagination is part and parcel of the emotional make-up is the fact that any emotion can be called up by images and by dreams. In the anticipation of a journey and its adventures we feel our hearts palpitating with a glow of excitement and happiness and we wake up in a cold sweat frightened by the images of a nightmare. A child can feel all the emotions of happiness and tragedy in the games and plays it produces through his imagination. In adults with a high degree of sensitivity and a strong imaginative factor images are constantly working on the mind and the imagination is the ruling power of their life. In those who are living more on an instinctual plane and whose psycho-motor reactions are more spontaneous, the imagination comes much less into play. The imaginative factor can be called the magic factor of the human mind. The imagination has an especially strong influence on sexual emotions and the sentiments of love, neither of which can be sustained without it. It is through our imagination that we experience any continuity of feelings. It is, unfortunately, beyond the frame of this book to go into further details about the power of the imagination and its grip on the emotions.

It is certainly not possible to judge the degree and intensity of imaginative powers from the inspection of a single part of the palm as chirologists want us to believe. One has to look at the whole hand, its general shape and size, the palmar skin, &c. There is, however, some direct significance in the ulnar part of the hand, particularly of the left. This can be explained through the following considerations: The function of the right hand, particularly of the radial part (thumb, index finger as well as the medius finger) is connected with the symbolic centres of the cortex, such as writing, reading, drawing, &c. One can call these the intellectual centres of the brain. The radial part of the hand is used in all manual activities and possesses the power of translating intellectual and motor impulses into action. The ulnar part of the hand has only an indirect significance in this respect. The left hand (in right-handers) has different functions altogether. The radial portion of the right hand is the most important tool of conscious activity and it is understandable from this that the quality of the intelligence has a special correlation with this part of the hand. The whole left hand as well as the ulnar part of the right have a different significance. The left is the passive side and not only in regard to the hand but to the whole body. I have written about this special aspect in my previous book and my views are supported by a number of authors from R. Fliess to Werner Wolff and others. One can assume that the whole left side of the body and the left hand in particular, are linked with the emotional and imaginative side of the personality and therefore with his subconscious mind. The ulnar part of the left hand ought to be especially significant in this respect as it is the least used for action. The same holds good to a minor degree for the ulnar zone of the right hand.

I shall now describe in detail where and how to find the potentialities of the imagination in hand-traits.

As always in any diagnosis from hand-features one has to proceed from the general to the single characteristics of the hand.

(*a*) THE GENERAL SHAPE

The predominant structural factor of the hand has to be diagnosed first. I say 'predominant', as every hand contains at least two different structural factors, sometimes more, but the one which is outstanding determines the hand type. One may thus find either elementary simple or regressive hands, fleshy motor and bony motor ones, and sensitive small or sensitive long ones. In people with sensitive long and sensitive small hands as well as in those with bony motor hands the imaginative factor is always a very strong one; the sensitive long hand has first place in this respect, then comes the sensitive small one and the bony motor types follow as slightly less characteristic of a highly developed imagination. For a detailed information see my book *The Human Hand.* The imaginative factor is interpreted from the knowledge of the constitutional make-up which goes with the different hand-types or, more precisely, with the different mixtures of hand-'elements'. The general shape of the hand allows therefore for a general and qualitative estimate of imaginative potentialities. Only in a general way can any quantitative assessment be made, namely according to the order into which I have put the different hand forms in their relation to introvert tendencies.

(*b*) ASYMMETRY OF RIGHT HAND AND LEFT

There are a number of hands with such a striking asymmetry that one represents a different type from the other. For example, the right hand of a person may be of a predominantly bony motor type and the left of a sensitive long, or the right of a fleshy motor form and the left of a sensitive small one. This description tallies with a good number of cases which I have investigated, and it must be remembered that there is a certain affinity between the asymmetric hands described. It is very rare that nature puts elements together which are structurally completely out of tune. In the two examples of left hand and right asymmetry (and there are many more possibilities) the left hand reveals an accentuated passivity not in itself as much as in comparison with the right. It is reasonable to assume from such a discrepancy that there is in the owner of such hands an unusually developed receptivity of the mind as well as a deep-seated constitutional conflict between his drives towards reality on one hand and imagination on the other. As the right hand is in both examples characteristic of a very strong active power which means physical dynamism, sportive activities and general participation in the world of action, the left hand shows a hypersensitivity towards impressions of both the outside world and the images of the mind, so much so that no inner harmony can be established. Because of constant frustration, the imaginative factor becomes more and more developed, often in a neurotic way.

(*c*) THE SIZE OF THE PALM IN PROPORTION TO THE FINGERS AND THE SIZE OF THE HYPOTHENAR EMINENCE

There are hands with long fingers and short palms and vice versa. An exaggerated example of the long-fingered type is the eunochoid hand and of the short-fingered one is the hand which goes with the Froehlich syndrome. It is also well known that the hands of monkeys and apes have short fingers and a long palm with a deep-seated and very prominent hypothenar eminence. The latter is much more developed in them than the thenar eminence as opposed to the condition in human beings. The hypothenar eminence of monkeys and apes is always demarcated by a deep crease, called the hypothenar line. It is common knowledge that monkeys and apes possess as a rule only one main transverse crease, while humans have two. When the simian condition is found in man, one speaks of a simian line, which is of genetic significance as it is an inherited feature. Now, not only the simian line but also a simian hypothenar eminence is present in a number of human hands, either together with the former or by itself. Both hand-traits therefore can be called atavistic. The simian line is most often found in a hand-type which resembles the elementary simple one and occurs together with other atavistic hand-traits; for this reason I have coined the term 'regressive elementary hand'. The over-developed hypothenar eminence is frequently present in the same type of hand but occurs as a rule in the sensitive small hand. Both varieties of hands possess most often short fingers, the elementary regressive one in about 70 per cent and the sensitive small in about 90 per cent of cases. The elementary regressive hand with an over-developed hypothenar eminence belongs to either mentally subnormal people or to abnormal personalities, whom one can call 'adventurers' either in body or mind. One finds the most daring soldiers with such a type of hand, people who are absolutely fearless and risk their lives without flinching. And adventurers of the mind, who strive for new roads in thought and art, are among those with elementary regressive hands. I only mention the hands of the composer Ravel and of the painter Balthus. According to my experience this hand-type goes not only with an unconventional mentality but also with a lack of emotional balance. The sensitive small hand goes with a versatile and at the same time infantile personality, which is highly imaginative, very insecure and often unstable. We know that atavistic attitudes are typical of many neurotics and that they are always accompanied by an intensified life of the imagination, often recalling that of native tribes. In such cases the imaginative factor is extremely pronounced. In people with a sensitive small hand the imaginative factor is constitutionally high and increases rather with age than diminishes because of persistent immaturity and insecurity. From this it is reasonable to assume that certain hand-forms, namely those with a long palm and short fingers as well as with an over-developed hypothenar eminence are characteristic of a high imaginative factor with a tendency to lack of proportion or even morbidity. A well-developed hypothenar eminence in other hand forms, such as the bony motor hand for example, reveals a high imaginative factor, which is more in tune with the whole of the personality than in the previous hand-types described. But the prevalence of either the constructive or destructive power of the imagination can only be judged from the ensemble of all hand-traits. One finds very rarely a strongly marked hypothenar eminence in the elementary simple and fleshy motor type of hand, both of which belong as I have repeatedly pointed out, to more cortically controlled personalities than the other hand-types.

The five fundamental factors of temperament which I have discussed in this chapter do not represent everything that there is to be said about the emotional make-up. They give, however, I think, a picture of the structural foundation of temperament and provide a good working basis for the diagnosis of the emotional make-up.

CHAPTER IV

THE HAND AND PERSONALITY

THE term personality is by no means a definite one. Different schools of psychology use it in different ways and give it different interpretations. The behaviourists believe that personality is mostly the product of environment, social psychologists are of the opinion that cultural patterns exercise an overwhelming influence on personality, and constitutional psychology sees personality mainly in the light of inborn tendencies. It is outside the frame of this book to discuss these different conceptions, but I must give my own definition of personality before I can start to show how personality is correlated with hand-features.

The very idea of personality seems to me to be that of differentiation. It is the opposite of 'type'. Individuality does not, however, comprise the whole of personality but it is the very essence of it. That which cannot be divided is the core or *gestalt* of the psyche and this individuality penetrates every action and reaction of a human being. It sets the pace, determines the rhythm and the curves of behaviour. It may be called the radioactive centre of the mind. The core of personality being an indivisible and individual entity cannot be further explained. For that reason we have to move, so to speak, from the centre to the periphery and to try to derive from the expressions of personality certain factors which represent its essential qualities. If we think of the whole personality as a circle, we can divide this circle into different sectors. In analysing the parts we shall of course not be able to find a synthesis of the whole, as the latter is something more: namely an organic entity which defies analysis. One can compare personality best with a work of art. Paintings by a great artist will always be recognizable among masses of other pictures because of their individual mark and style. There is a flavour and a something about them which is unique. A personality can have this mark and flavour of originality in different shades and degrees. And there is a first-rate quality about an individual only if he or she is really him—or herself. The absence of imitation and cliché is the first condition of being a personality. It is a strong Self and a good integration of both the subconscious and the conscious mind which seems to me to be the basis of a well-balanced and strong personality.

A personality with a strong ego will always refuse to ape others and this is what I mean by originality, which must not be confused with eccentricity. The strength of the ego depends on a number of factors both primary and subtle ones. Among the primary aspects of the ego are: the state of health, the instinctual drives and the amount of nervous energy. One often finds people with a weak vitality and a bad state of health possessing a strong ego, but these people build a strong ego up through some form or another of over-compensation or remarkable discipline; in the latter case this training of the ego is in itself a proof of its strength. The emotional make-up and its interaction with the self represents another 'sector' of the personality. The imagination and the intelligence constitute yet another one. The individual psycho-motor expressions, facial as well as verbal expressions, all these form part and parcel of the personality. The *kinetic*, the *emotional* and the *cognitive* factors as well as the *expressive style* all form sectors of the personality which are held together by the core or the Self.

It remains now to show if and how these different aspects are correlated with hand-features. I have emphasized in my definition of personality that its uniqueness or individuality is its very essence. And I have pointed out that individuality is the opposite to type. Why then do some psychologists, such as Jung and Kretschmer, speak of personality types? Jung's personality types, for example, are his main contribution to psychology. I think that Jung's term of personality types is a very unhappy choice but what he means is this: There are certain tendencies of personality which are common to large groups and because of these similarities psychology has accepted the term of personality types. This is in my view a misleading classification and I think those 'typical' qualities should be classified as general tendencies or general reactions types. Extraversion and introversion are certainly only general tendencies and it is the degree and intensity of these general tendencies which are strictly personal and different in every human being according to the whole of his psycho-physical organism.

Among other general tendencies than those just mentioned one can class the kinetic strength and emotional reaction types and as well as these two principal categories of the intellectual make-up, namely the more practical or concrete and the more symbolic or abstract intelligence.

There can be no question that these general tendencies can be diagnosed from hand-features, and seen at a glance, by the trained person.

(*a*) THE TENDENCIES TOWARDS EXTRAVERSION OR INTROVERSION

These are indicated by structural as well as by muscular hand-traits. The broad, short and muscular hand with strong and short fingers is typical of predominantly extravert tendencies. An exception to this is the elementary regressive hand, which fulfils up to a point the requirements just mentioned, but can easily be singled out through the following hand-traits: some atavistic features are always present, as for example a clubbed thumb, which is too short at the same time, the presence of the simian line and the absence of the long longitudinal line, apart from abnormalities of the papillary ridges on the thenar- and hypothenar eminences. The elementary regressive hand, the sensitive long and small hands—the former in particular—and the leptosome hand belong to people with predominantly introvert tendencies for reasons which I have already mentioned on several occasions.

(*b*) THE KINETIC STRENGTH

Kinetic strength is commonly called vitality. The physical must be distinguished from nervous vitality, which means that the potential dynamism is either used more for direct or instinctive action or for more indirect or planned action as well as thought processes. It is possible to diagnose from hand-traits which form of vitality is characteristic of the owner of the hands investigated. Broad and muscular hands with a clear and simple crease-line pattern belong to people with a strong physical dynamism and a good vitality. They are the men and women of direct action and are good athletes and sportsmen. They have, on the whole, natural attitudes to others and sound values. Remarkable nervous energy is often found in connexion with certain endocrine peculiarities—namely a slight degree of hyperthyroidism and in hyperadrenalism. In the latter case mainly women are affected this way. In constitutionally weak hands such as long sensitive hands the degree of both physical and nervous vitality is on the low side. In people with bony motor (leptosome) hands one often finds both strong nervous and physical vitality, but much of the nervous dynamism is wasted in fruitless battles of the mind, which fights to overcome the resistance and barrier of too much inhibition. The small, sensitive hand, especially of the broad variety (which looks like a fleshy motor hand in miniature), belongs to people of a mercurial make-up which goes with a very high amount of nervous energy, most often little balanced by the amount of physical energy available.

(*c*) THE EMOTIONAL MAKE-UP IN RELATION TO HAND-FEATURES has been dealt with in the previous chapter.

(*d*) THE INTELLIGENCE IN ITS VARIOUS ASPECTS AS SEEN IN THE HAND has been dealt with in the first chapter of this book.

(*e*) THE PSYCHO-MOTOR HABITS

Psycho-motor habits can be seen in their constitutional aspect in hand-traits, as I have already explained.

It is evident from this that the more primitive layers of personality, such as the kinetic and emotional factors, can easily be diagnosed from hand-features. In regard to intelligence, one can only give a general and qualitative estimate.

THE STRENGTH OF THE EGO

There remains only one more aspect to be discussed: is it possible to diagnose the strength of the ego and its integration or disintegration from hand-traits? The ego becomes apparent when the child starts to speak of himself as I. This is the birthday of his personality. The ego is built up by degrees through the kinaesthetic sense, and kinaesthesia is the opposite to the awareness of other people. Most gestures of the infant are limited to his own body. When the child explores his body by touching it, he achieves his first step towards the development into a personality. The evolution of personality goes from physical awareness towards mental realization. In this process *the hands* are the tools of the apprenticeship. The first autistic gestures so important for the mental development of the child, may reappear in later life and acquire then the opposite significance. People who suffer from insecurity and disintegration of the Self, regress to this type of gestures as a means of self-assurance.

This is sufficient to show how intimately the hand is linked with the ego. It remains to demonstrate now in which way the strength of the ego can be diagnosed from the hand. A strong ego is also a mature ego in the sense in which I use the word 'strong', and a weak ego is an immature one. The person with a weak or immature ego is naturally inclined to an easy disintegration. The type of hand which I have called the sensitive small hand is characteristic of emotional infantilism. One frequently finds in the sensitive small hand markedly bulbous pads on the finger-tips, which are characteristic of the infant and in an even more accentuated way of the embryo. The owner of such a hand might have a conspicuous personality, but his ego is immature and not stable. One can say, that all hands with infantile characteristics reveal immaturity and weakness of the ego. The reverse holds good for those hands which in their form and other characteristics are mature hands, provided that they show no signs of want of emotional balance as there are for example always in the elementary regressive hand. While it is a comparatively easy job to arrive at a diagnosis of the strength of the ego or its weakness from hand-features in the more pronounced cases, it is difficult to estimate the same in the so-called 'normal' person, by which I mean the man or woman of little individual colour. Normality represents in my view altogether a difficulty for psychological diagnosis. If one defines the 'normal' person as someone where no psychological difficulties exist, a failure in diagnosis need not be regretted too much. It is still possible to diagnose also in the normal person (if he or she exists) certain general tendencies, such as towards predominant extraversion or introversion, &c.

I must add at the end of this chapter, that the ego is, even where it is strong, a flexible entity and reacts to the experiences of life in undergoing changes. Those changes, which are less conditioned by the constitutional make-up than by the force of circumstances, must be measured by other tests than hand-diagnosis.

CHAPTER V

THE METHOD OF HAND-INTERPRETATION

A GENERAL OUTLINE

I HAVE given in my book *The Human Hand* a full description, covered by examples of my method of hand-interpretation. I must send the reader back to that book as a kind of preparatory course for this more advanced study. There are some general aspects in the similar chapter of *The Human Hand* which I no longer consider as valid in the light of the studies which I have since made and also because of my own development in psychological theory. I must also correct one misleading idea expressed in this earlier book, namely that the reader can learn the method of hand-interpretation from studying the book carefully. That is, I know, impossible, just as it is impossible to learn the diseases of the respiratory organs or any other organic systems from reading a book. Nobody can make a medical diagnosis after having studied medical literature and the same holds good for any study in psychology. That is why I have to warn the reader not to expect to gain any quick knowledge, not even if he or she is a fully fledged physician or psychologist. I am, however, convinced that the psychiatrist and the medical psychologist can acquire enough principal knowledge of my subject from reading this book to find out for themselves if the correlations between hand-traits and the different aspects of personality are applicable to the patients they are dealing with. And if they do not find them correct I shall be only too pleased to hear their criticism and to have perhaps been instrumental in provoking new research.

The reader who has perused *The Human Hand* will be familiar with the main six hand-types which I have established and their most frequent mixtures. I shall not repeat all that I have already said in earlier writings but I must give here a broad outline of the principal characteristics which go with the different hand-forms. While my classification into six main hand-types still holds good, my interpretation has changed in so far as certain correlations between the hand and temperament as well as mentality are concerned. I have omitted in this book any reference to the constitutional disposition to certain illnesses, a point which is stressed in *The Human Hand*. I have, on the other hand, emphasized much more the endocrine factor, because I have enlarged my knowledge of endocrinology through my research work under the supervision of Dr. S. L. Simpson. I have not treated the medical aspect of hand-diagnosis, first of all because this book is dealing with psychological diagnosis, and secondly because of the necessity to make far more studies on this special aspect than I have been able to do. The whole subject of hand-interpretation should be lined up with psycho-somatic medicine and the medical aspect is indeed, in a general way, part of it. I therefore hope to be able to enlarge my knowledge on this part of my subject in future research.

THE SIX FUNDAMENTAL HAND-TYPES

I have classified hands into two principal and functional groups, according to whether they are used predominantly for prehensile or tactile purposes, in the following way:

GROUP	TYPE OF HAND	SUBDIVISION
PREHENSILE	1. Elementary	simple
	2. Elementary	regressive
	3. Motor	fleshy
	4. Motor	bony
TACTILE	5. Sensitive	small
	6. Sensitive	long

I have already explained that the hand is an instrument of two types of function, namely (1) movements of purpose (cortical) and (2) movements of expression (subcortical). If the former prevail one finds the broad and muscular hand, if the latter prevail one finds the narrow and more or less flat hand. It is evident that the difference of hands in regard to their general shape goes with a difference of psycho-motor reactions in their owners. This fact has given me the idea of a new and simpler classification, built on the previous main six hand-types. I have divided hands into three groups, namely into broad, narrow and dysplastic ones. Groups 1 and 2 comprise all but the elementary regressive type of hand, which I have put into a group by itself together with all those hand-forms which are so to speak out of shape, which means cannot be classified. Needless to say that pure hand-types are the exception, that the mixture of two or three structural factors are the rule and that there are 'misfits' which defy every classification.

GROUP	SUBDIVISION	GENERAL ATTITUDES
A. BROAD	elementary simple hand fleshy motor ,,	predom. extravert tendencies ,, ,, ,,
B. NARROW	bony motor ,, sensitive long ,, sensitive small ,,	predom. introvert tendencies ,, ,, ,, ,, ,, ,,
C. DYSPLASTIC	elementary regressive and odd hands	predom. introvert tendencies

The reader will find among the illustrations both photographs of the back of the hands of the different types and prints of their palmar surface.

DESCRIPTION AND INTERPRETATION OF THE DIFFERENT HAND-TYPES

GROUP A: BROAD HANDS

Structural Traits

These two hand-forms (elementary simple and fleshy motor) are on a similar pattern. They are of medium length, broad and muscular. The outer edge is protruding, a formation for which the French have coined a very expressive term, namely *bord frappeur de la main.* The fingers are strong and as a rule rather stiff, especially in the elementary simple hand. They are short and broad, with short terminal phalanxes and furnished with nails mostly of medium size and broad. The palm is broad and thick because of the well-developed muscles which form the thenar- and hypothenar eminences. The latter is most developed in its distal part (the *bord frappeur*). *The skin* of the palm is rather coarse, especially in the elementary simple hand.

The Creases

The creases are deep, clearly designed and few in number. In the elementary hand one often finds only the four main creases. These are:

1. *The thenar line, demarcating the thenar eminence.*
2. *The upper transverse line, running from the 'bord frappeur' towards the index finger.*

3. *The lower transverse line, running in the opposite direction from the radial edge towards the hypothenar eminence.*
4. *The long longitudinal line, originating, as a rule, in the lower part of the palm and running towards the middle finger.*

The long longitudinal line divides, when well developed, the palm into two parts, a radial and an ulnar one. In broad hands like the elementary simple and the fleshy motor ones, the long longitudinal line is short and only crosses part of the palm; this is particularly so in the elementary simple hand. This crease is the most variable of the four main crease-lines.

One finds in the fleshy motor hand accessory creases, which are, as a rule, most concentrated on the thenar and hypothenar eminences, and run in a predominantly horizonal direction. There are few faults of design in the whole crease-line pattern of this type of hand.

The elementary simple hand has either no accessory creases or very few, running in horizontal direction, mostly situated on the thenar eminence.

Physical Qualities

Both types of hand go with either medium flexibility or rigidity, are of normal and healthy colour and on the whole warm and dry.

GENERAL INTERPRETATION

Both types of hands are found in people whose activities are mainly purposeful. The elementary simple hand belongs to individuals who work with their hands, such as peasants and manual labourers, gardeners, &c. They are also frequent among athletes, but with them the elementary regressive hand is also often present, probably a proof of a neurotic streak, often found in the professional heavyweight.

The fleshy motor hand which is much on the same pattern as the elementary simple one, differs from it in the less-marked rigidity of the fingers, a less coarse palmar skin and a more complex crease-line pattern. The fleshy motor hand is found in all walks of life with people of a certain robustness. They are men or women of action in whom reason and the sense of purpose have first place: cortical impulses outweigh subcortical reactions. Such people show a natural simplicity, which goes with an easy integration. In them spatial and practical intelligence are far more developed than symbolical and abstract thought. They always want and need an outdoor life, much movement and fresh air. This type of person is generally good with his hands and knows how to use tools. As I have already said, boxers, wrestlers and football players have often such a hand. It is noteworthy that musicians, particularly pianists, possess often fleshy motor or elementary hands. The popularly called 'artistic' hand, which corresponds to the long sensitive hand-type, is rarely found in artists who use their hands, especially in men. The idea of the 'artistic' hand is a hangover of a certain fashion in painting when the long narrow and flexible hands were thought of as cultured and refined and—artistic. People with elementary simple and fleshy motor hands are of a powerful physique; they are inclined to get stout round the waist in middle age and according to Sheldon's somatotypes they are predominantly endomorph-mesomorph. In the case of the elementary simple hand they are predominantly mesomorph-endomorph. The digestive and muscular apparatus have functional and psychological predominance. One can describe such personalities as primitive. This primitivity, by which I mean simplicity, accounts for their generally good mental and emotional balance, their zest for life and a naturally benevolent attitude towards their fellowmen. They are people with predominantly extravert tendencies.

The endocrine factor is characterized by a tendency to hypothyroidism.

MAIN CHARACTERISTICS OF PERSONALITY

Vitality

The owners of both types of hands have constitutionally more physical than nervous energy; this is most definitely the case with people who possess elementary simple hands. With fleshy motor hands one frequently finds both strong physical and nervous vitality. Strong instinctual drives are characteristic of the owners of both hand-types.

Emotional Make-Up

Together with strong instinctual drives one finds especially in the owner of the fleshy motor hand a marked sensuousness; love of good food and other sensual pleasures. The emotional reactions are on the slow side, and in the more primitive person with elementary simple hands, the temperament may be called phlegmatic. In him sensitivity is rather on a restricted scale. While the same tendencies are present in the far more complex owner of the fleshy motor hand, both the speed of reactions and the range of sensitivity as well as its acuteness, become more pronounced.

There is often a competitive spirit in the latter type of personality; but he or she is, by nature, capable of 'fair play'. The desire to compete is not pronounced or absent altogether in the owner of the elementary simple hand, who is, as a rule, a placid person. In both groups of people, *patriarchal attitudes* are the rule, with a genuine desire to protect and help the weak. The imaginative factor is generally not much developed. In any case, it rarely takes a prominent place.

Intelligence

The intelligence differs very widely both in quality and range, but it is always on the side of predominantly practical and concrete thinking.

Capacity for Integration

This is especially high in both groups of persons.

GROUP B : NARROW HANDS

1. THE BONY MOTOR OR LEPTOSOME HAND

This hand is very frequent among the English. It belongs to tall, slim and muscular people. The golf player, the good rider and swimmer, are among the leptosome variety of human beings. According to Sheldon, the physical type just described possesses both a strong mesomorphic and ectomorphic factor. The reader must not get the impression that the leptosome person is destined to be a sportsman only, or the pyknic person an athlete. One finds all physical types in practically all walks of life and in most professions, except those which demand a special psycho-physical make-up, as for example the profession of the dancer. Thus one meets the owner of the bony motor hand among sportsmen, intellectuals, artists and business men. Great surgeons are frequently leptosome people with bony motor hands.

The Structure of the Hand

This is, as the name indicates, a hand with a strong bony structure, with a long palm and long fingers. The palm is on the narrow side, as the muscles, though well developed, are not particularly voluminous. The distal part of the palm is flat. There is a much less developed *bord frappeur* than in the fleshy motor hand, but the hypothenar eminence is generally very marked in its lower part, and often deep seated as well.

The fingers are strong and long and rather rigid as a rule. The nails are mostly of a longish shape, either narrow or broad, having well-developed or very large moons.

The Palmar Skin. The skin is less coarse than in the broad hands described.

The Creases. The crease-line system is complex. The flexure lines are deep and numerous. The accessory lines run in both the horizontal and longitudinal direction. The long longitudinal line, which is so often undeveloped in the two previous hand-types described, is always well marked.

The design of the main creases is most often hesitant and faulty.

The Papillary Ridges

In a certain number of bony motor hands, unusual patterns of the papillary ridges are present on the thenar- and hypothenar eminences, such as loops and whorls.

Physical Qualities

The bony motor hand is always a stiff hand. Other physical qualities vary too much to be described as typical.

GENERAL INTERPRETATION

This is a complex hand which is found in people with what is popularly called a 'divided' mind. A good vitality and strong nervous dynamism make them enjoy outdoor activities and sports, but a purely rustic life is for the owner of the leptosome hand of the intellectual variety, not enough of an outlet. There is a great drive towards intellectual pursuits and interests of a contemplative and aesthetic nature. One can describe the character of the leptosome man as that of the typical escapist, a man who loves to live in a kind of ivory tower in the country, far away from the humdrum of daily life. On the other hand, a strong dynamism and sportive trends drive these people towards social and human contacts, but there is always some tendency towards inner isolation. Human relationships are difficult for them. There is something hard and angular in their physique as well as in their emotional make-up. They tend to be obsessional and pedantic. Their appearance is *soigné*, often meticulously so, and they will insist on having everything put in its right place.

MAIN PERSONALITY CHARACTERISTICS

Predominantly introvert tendencies.

Vitality

Very good vitality and strong drives, both physical and mental; nervous energy is as a rule stronger than physical energy. Mostly very tenacious people with a great sense of endurance.

Emotional Make-Up

The leptosome person has a tendency to hyperthyroidism, which goes with nervous restlessness and little power of concentration. The psycho-motor reactions are characterized by quickness of movements and manual skill in regard to fine and subtle movements of the fingers, which are the natural equipment for fine crafts and arts.

The marked emotional inhibitions of this type of person lead to impulsiveness and an abrupt manner. A generally high nervous tension produces stiffness and reserve in human contacts. Receptivity is highly developed, which goes with impressionability towards the outer world, and also in regard for the atmosphere produced by human beings, towards the beauty of nature and the world of art. Aesthetic appreciation is great and often overvalued. The owner of the bony motor hand is a fastidious person and difficult to please in every respect. That is why he gets easily disappointed in others and also in himself. His human contacts being difficult, he tries to escape into the world of books or solitude in nature. His craving for intellectual contact and stimulation throws him always back into the society of men, until he gets dissatisfied again and is thrown back on himself. This is the background of a perpetual emotional conflict for which it is difficult to find a solution. Physical vitality and sportive activities provide a compensation and assure a reasonable amount of health. The imagination is much developed.

The Intelligence is of a predominantly abstract type. One finds great statesmen and military leaders among the owners of the bony motor hand.

Capacity for Integration

It follows from the description of the emotional make-up that the capacity for integration is low in this group of people, who need a stereotyped kind of life with much routine, in order to keep a well-preserved façade and some real balance.

2. THE SENSITIVE LONG HAND

This is a weak hand, especially when found in a man. This type of hand is called 'artistic' because of its aesthetic appeal. As I have already pointed out, the Renaissance painters have contributed to the 'idealistic' interpretation of the long, sensitive hand by the ignorant. Botticelli, for example, took this type of hand as a model for all his Madonnas, and Leonardo da Vinci's famous Mona Lisa still bewitches her admirers with her beautiful long, sensitive hands, which are very weak from a psychological point of view.

The Structure

All that I have said in the chapter 'The Endocrine Factor' about the eunochoid hand holds good to a minor degree, for the long sensitive hand, the former being an exaggerated form of the latter, which is very long and narrow, of delicate bony structure and poor muscular development. The fingers are very long, thin, and mostly tapered, furnished with long and narrow nails.

The Skin is very fine and often glossy.

The Creases. The flexure lines are superficially drawn, and there are many faults in the main creases in the majority of cases I have seen. The accessory creases are also very fine and very numerous, running in both longitudinal and horizontal direction.

The Physical Qualities. The fingers are always very flexible, the tips can be bent backward in most cases. The flexibility of the palm varies, but is as a rule, high. The hands are often pale in colour and cold to touch.

GENERAL INTERPRETATION

The owner of this hand has all the qualities which I have outlined under the heading 'The Eunochoid Hand' in Chapter II. I shall therefore not repeat this description here, but only indicate the principal characteristics of personality.

CHARACTERISTICS OF PERSONALITY AND INTELLIGENCE

The owner of the long sensitive hand has poor drives and a low vitality. His endocrine factor shows signs of subgonadism. There is always marked tendency towards introversion. The emotions are of a shallow nature and on a narrow scale because of a strong inhibitive factor, which prevents their free expression. The whole inner life is more imaginative than in any other group of human beings. The intelligence is often exceptionally high and definitely much more prone towards abstract than concrete thought. There is often a tendency towards mysticism and occultism. The capacity for integration is, as one must expect from what I have just described, particularly low.

3. THE SENSITIVE SMALL HAND

This type resembles in many respects the 'infantile hand' which I have described in the chapter 'The Endocrine Factor', and there is probably a tendency to hypopituitary infantilism in its owner.

The Structure of the Hand

This is, as its name implies, a small hand and found in people of small stature. It is a woman's hand, and its appearance in men always designates feminine traits of mentality and temperament. The bony structure is delicate but less so than in the sensitive long hand. There are two varieties of the sensitive small hand, a narrow and a broad one. The bones of the latter are stronger and the hand as a whole resembles the fleshy motor hand in miniature. The palm is long and the fingers short, mostly with square, and sometimes with tapered finger-tips. The latter are furnished with nails of medium size, mainly broad and with good and visible moons. A characteristic feature is the elongated hypothenar eminence, which resembles that found in monkeys and apes. The thumb is on the small side and often double jointed.

The Skin. The palmar skin is soft and delicate and often glossy.

The Creases. In the narrow variety of the small sensitive hand the creases are very numerous and superficially drawn, the palmar crease pattern resembling that of the long, sensitive hand. In the broad variety one finds very well drawn and deep creases similar to those of the fleshy motor hand, but the accessory creases are more numerous and of a more superficial design. They are specially marked on the distal part of the palm; which has most often slightly padded eminences beneath the third and fourth fingers with papillary ridges in forms of whorls or loops.

The Physical Qualities. Flexibility varies widely, but generally this is a flexible hand. The fingers and especially the thumb are often double jointed. I found more often warm and dry hands than the opposite among this group. The colour, on the whole, is pink or pale.

I am breaking away here from the method applied in describing the different forms of hands and am quoting mainly all that I have said about Constitution, Temperament and Mentality as well as Vocation in *The Human Hand* as this description seems to me still the best I can supply.

Constitution

As one would expect, a small physique and a delicate and nervous constitution go with this hand. Endocrine functioning is marked by a tendency towards hypopituitary infantilism. The autonomic nervous system is often hypersensitive, and this is frequently the cause of allergic illnesses such as asthma or nervous colitis. The pronounced flexibility of the hand reveals a general astheny which goes, as a rule, with lack of energy. Lack of vitality and a highly strung disposition prepare the way for those nervous disorders of which the neurasthenic person is so often a victim.

Temperament

His hypersensitivity and excitable nature deprive him of emotional balance. His emotions are easily roused but his resistance and endurance are minimal. He reacts compliantly to impressions and becomes easily a prey to his surroundings. A calm and harmonious atmosphere is indispensable to his health and mental equilibrium. In moments of real danger, however, he suddenly shows vigour and surprises one by his sense of responsibility and resources of unexpected strength. It is as if his vitality consolidates itself under external pressure. He retains childish features all his life and, being unstable and egocentric, a subjective world of romantic design offers some form of escape from the harshness and injuries of reality.

Mentality

His intelligence is vivid and original. The neurasthenic type of man furnishes the world with a great number of its intellectuals. Open to all impressions he works on them with intuition and imagination and is thus enabled to understand others. In this group one finds people whose vocabulary is rich and original and who have a good literary style and a strongly developed critical sense.

Vocation

The ancients called the 'nervous' man the Mercury type. His character and mentality being very flexible and indefinite, his talents are varied and they are on a considerable scale. A representative member of the intellectual and artist classes, he is naturally gifted for literature and journalism and is often a good actor.

But the vocation of people with small infantile hands is not confined to brain work; we find them in such professions as fashion designing and interior decorating.

GROUP C: DYSPLASTIC HANDS

1. THE ELEMENTARY REGRESSIVE HAND

I have classed in a group by itself the elementary regressive hand, which is the most puzzling of all. It is owned by feeble-minded subjects as well as highly intelligent people. I have chosen the term 'elementary regressive' because of the similarity of this hand to the elementary simple hand in regard to shape and general characteristics of the crease-lines. What distinguishes the one from the other is the presence of one or more 'simian' features in the elementary regressive hand, hence the designation 'regressive'. One finds either the simian line or a very small and ugly thumb with a short and clubbed terminal phalanx and an unusually small fifth finger. The patterns of the papillary ridges also very frequently resemble those found in monkeys and apes. Another characteristic feature, present in a high percentage of cases is the absence of the long longitudinal line. All these traits are missing in the elementary simple hand. And so is the pronounced flatness of the distal part of the palm, characteristic of the elementary regressive hand, which is generally found in the bony motor type of hand. I am demonstrating in the clinical studies reported in the second part of this book that the elementary regressive hand is found in a very high percentage in feeble-minded people with or without character disturbances. Atavistic hand-features increase down the scale of mental deficiency either in this type of hand or in any other.

The astonishing fact that the elementary regressive hand is found in not a small number of cases in highly intelligent but neurotic individuals defies any explanation and must be accepted as a matter of experience. I have never met with this hand-form in any well-balanced personality.

I have seen it in a number of artists, particularly musicians and painters, and also in some scientists, all of whom suffered from severe neurotic disturbances. There were two among them who have achieved outstanding fame.

The Structure of the Hand

Some characteristic structural features have been already described in the previous paragraph and I shall not repeat them here. The elementary regressive hand is of medium length, with either short or long fingers, which are less stout than in the elementary simple hand. They are often very rigid and 'inarticulate', a typical feature of the hands of mental defectives. The nails are characteristic, too. They are always very short and may be called 'primary'. One finds moons very rarely. Concave nails or other 'pathological' shapes are frequent.

The hand is very broad in the lower part of the palm, while the upper part narrows considerably and is flat. The thenar eminence is mostly well developed and the hypothenar eminence often very pronounced and deep seated, a feature which is also uncommon in the elementary simple hand.

The Palmar Skin. In those elementary regressive hands which belong to feeble-minded people the skin is coarse and the papillary ridges are very widely spaced. In those which one finds in the intelligent type of person these two characteristics are missing.

The Creases. In the case of the feeble-minded with elementary regressive hands there are only the main creases present, and these are often abnormal, either through the presence of the simian line or the absence of the long longitudinal line. One finds thus hands with only two main creases and no accessory ones at all. If there are, they are very few in number and mostly situated on the thenar eminence and on the distal part of the palm. The lines are very deeply marked and 'clumsily' drawn. The situation is different in the case of the highly intelligent person. With him the creases are more numerous and better in design, while there is an equal frequency in the presence of the simian line or absence of the long longitudinal line.

The Papillary Ridges. They deserve a special mention in elementary regressive hands. Loops and whorls on the thenar and hypothenar eminences are frequent and the simple triangular pattern appears on the finger-tips, either on one finger or on several. I have seen cases where all fingers were affected.

The Physical Qualities. Flexibility of both the palm and the fingers is always low. The other physical qualities vary too widely to allow for generalizations.

GENERAL INTERPRETATION

The elementary regressive hand is that of a throwback. Regression into an earlier form of existence, both mentally and emotionally, is the common denominator of a hand-form of such widely different significance. Whatever the intellectual level may be, introvert tendencies always outweigh extravert ones. In both, the person with poor or with unusual intelligence one finds a poor social adaptation and little emotional equilibrium. The feeble-minded individual with this type of hand is prone to violence or at least compulsive reactions of an aggressive nature, his opposite number on the intellectual scale tends to an outright rebellious attitude and to obsessional reactions in his personal contacts. Destructive impulses directed towards the Self as well as towards others make these people 'the problem' of their milieu. It is difficult to cope with them, and it is difficult for them to cope with themselves. This sounds like a completely negative verdict on any owner of the elementary regressive hand. One must, however, not forget that the gifted neurotic of this category possesses often an unusual creative power and great richness and originality of the imagination. He is the true adventurer of the mind. I must make it clear that among the intelligent owners of this hand there are people of all walks of life. They are reckless people, who love the unusual and the unconventional and who make excellent soldiers and leaders of men, because of their daring spirit and sense of self-sacrifice. While they are developing 'virtues' far above the average person in unusual circumstances they are very easily demoralized in the humdrum of every day life. Their greatest difficulty is their lack of discipline.

I am omitting a description of characteristics, of personality and intelligence as they are mostly covered by what I have said in the general interpretation.

Capacity for Integration

As is obvious from the previous text, the capacity for integration is especially low. Indeed one can say that an elementary regressive hand goes always with a constitutionally disintegrated personality.

2. ODD HANDS

One sees often in imbeciles and idiots acromicry, because of their thwarted growth which frequently amounts to dwarfism. Acromicric hands and all those which are a misfit and cannot be classed into any of the groups expounded, I have called 'odd hands'.

This description of the six fundamental hand-types is meant as a guide for the classification of hands according to their structural and other characteristics, but I must make it quite clear that there are no pure hand-types to be put safely into pigeonholes. Every pair of hands consists of a mixture of hand-forms and it is essential to discriminate clearly all the different elements involved. It is also essential to realize the degree and the intensity of each hand element in the ensemble of hand-traits. One could speak of hand-factors as Sheldon speaks of 'morphological' factors the intensity of which allows for the diagnosis of the psycho-physical make-up as a whole. Hand-diagnosis shows that one part of the body can be used for the psycho-biological interpretation of the whole personality. Why the hand is so particularly suited for a special and specialized psycho-physical study of man I have tried to explain in practically all my writings. It is my conviction for which I have given some theoretical and much practical evidence that there is no part of the body which can tell us more about personality than the hand. The concentrated study of one part of the body which can tell 'the whole story' seems to me to be preferable to the study of the whole body for the same purpose, both for reasons of time and the avoidance of the tiresome proceeding of a wholesale anthropometric investigation, a proceeding which a psychologist could not undertake on his patients for obvious reasons. And it is particularly the psychologist who should use a psycho-physical test method. The study of the hand also entails a wider range of psychological interpretations than purely anthropometric data can offer which puts hand-diagnosis in a class by itself.

RIGHT HAND AND LEFT

There are no pairs of human hands alike and the right and the left hand of the same person are never on the same pattern. I have already mentioned in this book that there is not only a division of function in the right hand and left but also a difference of psychological significance. The psychological meaning of the right and left sides of the whole body and of the face in particular has attracted the attention of several authors among whom I only mention Werner

Wolff, who in his book *Expressions of Personality* shows so convincingly the physiognomical and psychological split of the two halves of the face. I have shown in *The Human Hand* that the right and left hand are never completely symmetric and that the left palm is in the majority of cases longer than the right; the condition is reversed in regard to finger length. I have correlated these structural differences with differences in psychological significance and have given other evidence about the same subject in my book *A Psychology of Gesture* through a study of hand-gestures. In a recent number of *Enfance*, in an article which I have already mentioned, Roudenescu and Theys give proof on similar lines to myself of the difference in psychological significance of the right hand and left.

I have not changed in any way my ideas about this special aspect of my subject since I wrote *The Human Hand*, and for this reason I am not giving a full discussion about the psychological significance of the right hand and left here, but must send the reader back to my earlier book, where he will find the information required.

THE GENETIC SIGNIFICANCE

Some work has been done on the genetic significance of hand-traits, especially in America. The investigations of D. Rife, for example (1941 and 1943), were concerned with 'Palm Patterns and Handedness' and with the 'Genetic Interrelationships of Dermatoglyphics and Functional Handedness'. Leche and Newman were dealing with the same problems. All three investigators agreed that special patterns of the papillary ridges on the thenar- and hypothenar eminences are more frequent in lefthanders than in righthanders. And the same holds good for unusual patterns of the papillary ridges in the interdigital areas. These are very specialized studies and limited to the papillary ridges only, but they prove without a doubt the genetic significance of hand-features.

In this country S. L. Penrose has made a special study of the genetic significance of the hand, but has confined himself so far to the investigation of the papillary ridges and lefthandedness. As everybody knows, a whole systematic science has been developed on the significance of the papillary ridges of the finger-tips for the purpose of identification. This needs no further mention. It is however important to realize that it took scientists over half a century to advance from the 'Fingerprints' and their genetic significance to the idea that similar patterns on other parts of the hand may also indicate something of scientific value. The fact that lefthandedness too has for some time attracted the attention of both the educational psychologist and the geneticist can also be regarded as a step forward towards the wider comprehension of the significance of hand-traits. There is no literature at all, according to my knowledge, about the genetic significance of single crease-lines. It is well known that left-handedness is hereditary, and, as I have been able to show, this characteristic goes often with the presence of the simian line, which also runs in families. The simian line mostly occurs in the elementary regressive type of hand, where one also finds most often unusual patterns of the papillary ridges on the thenar- and hypothenar eminences. This hand, which is significantly frequent in mental defectives deserves, in my opinion, a special attention in regard to genetic studies.

Genetics are not my subject and I thus conclude this short paragraph with the suggestion to workers in this special field to broaden their study of the hand and to look for possible correlations between certain hand-forms, as for example the elementary regressive hand and single hand features which are genetically characteristic.

THE NAILS

The nails are mainly characteristic of health conditions. They are the only part of the hand which renews itself regularly and constantly and reflects therefore, just as the hair does, the state of vitality and certain health conditions. The nails in heart disease, in tuberculosis and other illnesses of the respiratory organs are well known, but there is a lot more to learn about the correlation between the nails and disease. Nail formation has also some genetic significance as Padro Castello has been able to show and there is a connexion between the insertion and the shape of the nails in mental deficiency, as I have demonstrated in my studies on mentally defective children and adults. As I am not dealing in this book with the medical aspect of hand-traits except when it is directly related to the psychological one, I must send the reader again back to my book *The Human Hand* where he finds a more detailed information about this subject.

THE LINES OF THE HAND

There are two kinds of lines on palm and fingers, namely the papillary ridges and the crease-lines. The former cover the hand in evenly spaced ridges, running in different directions and forming either triangular forms or arches, whorls and loops on the finger-tips. Whorls and loops also appear in the interdigital spaces, more often on the eminence underneath the third and fourth finger than on that underneath the second one. They increase in frequency in the direction of the ulnar edge of the palm, and are comparatively rare on the thenar- and hypothenar eminences, as I have already mentioned earlier in this chapter. The design of these patterns is of such individual variety that they are never the same in two different pairs of hands, and not even on the right and left hand of the same person. This individuality as well as the fact that they are already formed in prenatal life and never change from birth to death has been used for the ingenious device of identification from finger-prints. Almost all the scientific literature on hand-traits from Purkinje to H. Cummins and D. Rife deals with these dermatoglyphics. Harold Cummins, of Tulane University, New Orleans, specially investigated palmar patterns of the papillary ridges in different races and found significant differences in their design and position in Jews, European Americans and negroes from West Africa. He also pointed out that there is a correlation between palmar patterns of papillary ridges and the simian line in mongolian imbeciles. He also mentions, that with them the papillary ridges on the finger-tips show a remarkable reduction in the occurrence of whorls. I have confirmed these findings through my own studies on the hand of mental defectives and have been able to show that the same correlation occurs also in other types of mental defectives than mongolian imbeciles, and, what seems of importance too, the hands of normally or highly intelligent people with a simian line present the same condition. Cummins also observed that hypothenar, second and third interdigital patterns are characteristically more abundant in right hands in the ordinary population than in the left. In mongolian imbeciles the dextral accentuation is lessened as well as the pronounced differentiation of the right hand and left altogether. D. Rife concentrated his researches on the correlation between handedness and papillary patterns and demonstrated the genetic significance of both. These examples suffice to show the importance of systematic research of the papillary ridges for constitutional psychology. I may add that I have observed a correlation between the different spacing of the papillary ridges in different hand-types. In the elementary types of hand the spacing is wider than in the sensitive types of hand. There is, indeed, a lot more to be learnt about both kinds of lines, the papillary ridges and the creases, than science has hitherto admitted, in spite of the fact that some enlightened pioneers, such as Purkinje, Bell, Wood Jones and others have recommended the study of the hand to enlarge our knowledge about the 'nature' of man.

More than a century ago (1823) Purkinje published a thesis, entitled 'Commentatio de examine physiologico organo visus et systematis cutanei', a thesis which has been examined and made accessible for the 'ordinary' reader by H. Cummins and Rebecca Wright Kennedy in a paper 'Purkinje's Observations on Finger Prints and Other Skin Features'. One finds the following statements in Purkinje's thesis as reported by the above-mentioned authors:

> Upon tactile examination the skin is hard or soft, moist oily, clam or dry, warm or cold, elastic, rigid or spongy smooth or rough, loose or taut, pliant or non pliant. These qualities are found in different ways and degrees or variously modified in different states of vital activity. It is to these properties that attention must be directed, for the revealing of differences, however slight, in temperament, constitution, age, sex, or even family and nationality.

and, in regard to papillary ridges, Purkinje says:

> Our attention is now attracted by the remarkable disposition and flexures of the continuous rugae and sulci in the inner part of the hand and foot, especially in the terminal phalanges of the fingers. In general, indeed, some mention of them occurs in every physiological and anatomical epitome. But in an organ of such importance as the human hand, which serves not only for movement of many different types but also for the sense of touch, every inquiry, however minute, brings to light some worthwhile information.

THE CREASE-LINES

There are four main crease-lines:

1. *The thenar line, which demarcates the thenar eminence.*
2. *The upper transverse line, coming from the ulnar edge of the palm and ending either beneath the index finger or in the interdigital space.*
3. *The lower transverse line, starting from the radial edge of the palm and ending most often on the upper part of the hypothenar eminence. This line is more variable than the previous one in both its design and length.*
4. *The Long Longitudinal Line. This is the least stable crease-line of the four. When it is fully developed, it starts on the base of the palm, running obliquely across the palm towards the second interdigital space. But in a great number of hands only a portion of the long longitudinal line exists and in quite a number of palms it is missing altogether.*

The creases are related to the joints of the hand and follow their curves. They are folds of the skin, which are formed through the movements of the hand and fingers. While the transverse creases are due to the movements of the four fingers, the thenar line owes its existence to movements of the thumb. The long longitudinal line is due to movements of opposition of the thumb as well as of the fifth finger. The particular mobility of the human hand is achieved through the freedom of movement of each finger, and especially of the thumb and index finger. Man only is able to move the two latter fingers separately, a faculty which essentially differentiates the human from the simian hand and produces the main two transverse lines instead of one in the hands of monkeys and apes. In them one transverse crease runs straight across the palm from the outer to the inner edge. In human beings the main transverse creases, especially the lower one, are curved and show a great individual variety in their length, their position and the élan of their design. The thenar line as well as the main transverse creases are already formed in prenatal life and generally do not change during lifetime. Not so the long longitudinal line, which develops most often after birth during the first years of life.

THE ACCESSORY CREASES

While the principal creases are of a more or less stable design and position on the palm, the accessory creases show such an individual variety that it is only possible to classify them in a general way, namely according to their position and 'intensity' on certain parts of the palm. There are, however, two accessory creases which are more typical than the rest and which appear always on the same place of the palm. These lines are:

1. The hypothenar line, which demarcates the hypothenar eminence, and may be called a regressive hand-trait as common to monkeys and apes. In humans this crease is present, as a rule, in the small sensitive hand, which possesses an 'overdeveloped' hypothenar eminence. But one finds also in mixed hand-forms the same condition. The hypothenar line is practically never present in elementary simple and fleshy motor hands.

2. The distal arc is a semi-circular crease-line at the base of the fingers, known in chiromancy as the girdle of Venus. It is due to separate movements of the third and fourth fingers and it occurs most often in specially mobile hands. It is usual in the sensitive small and long hands, and sometimes present in the bony motor and

elementary regressive hands. It is practically never found in the elementary simple and fleshy motor hands.

The Origin of the Accessory Creases

The accessory lines comprise two groups: (1) A number of small creases, which arise wherever the skin folds as a consequence of hand movements and which, though related to movements, are more indirectly so than the main flexure lines. (2) The rest of the accessory lines, for which the explanation of purposeful movements is insufficient. No scientist has to my knowledge clearly differentiated the latter from the previous ones in regard to their origin and significance. As they are of the greatest importance for my subject, I shall give them fuller treatment.

One must distinguish voluntary and more or less purposeful movements from involuntary and more or less expressive ones. While the first are related to conscious impulse and thought, the latter can be generally regarded as reflex movements which are related to nervous and emotional impulses—those largely emanating from the subconscious mind. That a certain number of accessory crease-lines are due more to involuntary than to voluntary hand-movements will be understood by the following:

The accessory crease-lines are not in proportion to the manual activity of the hand-owner. On the contrary, lazy hands are the richest in those creases. The hands of intellectual and nervous people—sensitive and to a certain extent motor bony hands—are little used for manual work, yet these are the ones in which we find close systems of accessory lines. It can be argued that in them the underlying tissues are looser than in the elementary simple and motor fleshy hands of manually active people; but this is only half the truth. The more sensitive a person is, the greater is his need for self-expression and the richer his natural vocabulary of gesture. Gesture involves complex idiosyncratic movements necessitating a high degree of mobility of the hand and causing many accessory creases. The meaning of gesture is wholly psychological. It is profoundly allied with nervous and emotional reactions, and only to a certain degree with stimuli of thought.

In a statistical survey of 1,600 hands, all belonging to right-handed people, I found more accessory crease-lines in the left than in the right hand. No mechanical explanation can reasonably be given for this. It is a fact that illustrates beyond doubt that the crease-lines of this type register other than simple purpose movements.

The more frequent and complex a person's nervous and emotional stimuli, the more will emotional tension and nervousness be produced in him, and this will affect the scale and variety of his involuntary movements and gestures and will be registered in the accessory crease-lines of the palm. We have, therefore, a correlation between the quantity and quality of the accessory creases and the nervous make-up, for which I found also negative evidence. In thirty-two hands of boxers and wrestlers, all of the elementary type, there was in every case a minimum of accessory crease-lines. I found the same in the hands of feeble-minded people and high-grade imbeciles at an L.C.C. mental hospital. Dr. Earl has pointed out to me that he has observed among high-grade imbeciles in his particular hospital 'the boxer's hand' and also a certain abnormality of creases in many cases 'where no other dysplasia could be found'. Besides the experimental evidence there is theoretical evidence. It relates to the origin of these lines and discards the conventional idea that they are all folds of the skin originating through purpose movements.

On every palm bearing complex patterns of the papillary ridges, such as loops and whorls on the thenar and hypothenar eminences, one finds accumulations of small crease-lines running centripetally towards them. I have already explained that these formations of the papillary ridges mark centres of the tactile sense, and the crease-lines running towards them might have something to do with emotional or nervous disturbances; an idea which is supported by the fact that both these patterns and these creases appear in the majority of cases in the hands of abnormal personalities.

It is clear from this that:

1. *The accessory crease-lines are psychologically more interesting than the principal creases.*

2. *There is little significance in the singular lines, but in the quality and quantity of the accessory creases as a whole.*

The quality of the skin is also a factor for the quantity and quality of the creases. In a hand with a coarse palmar skin one finds few of them, or none; in a hand with a delicate palmar surface the position is reversed. The texture of the palmar skin is part of the ectomorph factor of the whole constitutional make-up. One always finds syndromes or groups of traits correlated with different forms of hands; for example a coarse skin, few accessory crease-lines, deep and clearly drawn main crease-lines go with the elementary type of hand. A fine skin, many accessory creases, superficially drawn main creases go with the sensitive types of hand, &c. Several syndromes, sometimes of an opposite kind, may overlap in one and the same hand, if the latter is a mixture of largely diverging structural elements. The more complex the psycho-biological make-up of the person, the more so will this be the case. There appears to be however a law which limits the possibility of structural combinations of hand elements as well as the syndromes belonging to them, as certain structural elements exclude others.

HAND MEASUREMENTS

Hand measurements are an essential part of the scientific investigation of the hand. Anthropometry has been both over- and undervalued by scientists. The possibility to decipher man's personality from his bodily proportions would be an ideal method. Needless to say, I do not believe that such a method could ever lead to any fruitful results. An all-round study of the mind cannot be attempted by physical measurements only, as this would be a contradiction or at least a profound misunderstanding of both: the constitutional factor and the mind. The latter is far too intangible and fluid, so to speak, to be put into the fixed scale of measurements. Measurements of the body in general, and of the hand in particular, are only useful up to a point; they teach us something about different elements or factors (ectomorph, mesomorph and endomorph according to Sheldon) which compose human physique. They also reveal a lot about the endocrine make-up. As far as the interpretation of the individual mixture of the three factors and of the endocrine formula allow for psychological conclusions without subjective additions, one can accept their value for constitutional psychology.

The estimation of psychological data from hand measurements is limited but is as important as any other application of anthropometry in the sense of Kretschmer and Sheldon. Naturally, hand measurements have to be compared with other physical measurements, such as height and weight of the person concerned, as they are only of relative and not of absolute value. It is nonsensical, for example, to speak of a long hand without comparing it with the height of the person, &c. While the length of the hand must be compared with the height, the breadth must be compared with the weight of the person. A broad hand in a person with very slender physique has a significance different from that of the same hand in a stout person.

Apart from comparing the length and width of the hand with height and weight, the proportions of the different parts of the hand must be compared with each other. It is thus necessary to investigate:

1. The length of the whole hand, which is measured from the middle of the wrist, half way between the processus ulnae and radii, to the tip of the medius finger.
2. The length of the palm, which is measured from the same starting-point as the length of the whole hand, to the articulation of the third metacarpal bone with the medius finger.
3. The length of each individual finger.
4. The length of the phalanxes of the fingers.

The fingers should be measured from both the dorsal and the palmar side. In the latter case, one is able to judge the length of the palm in relation to finger length. In a number of cases the palmar skin covers more of the basal phalanxes than in others, and in most cases the palmar surface is longer in the left hand than in the right.

The width of the hand is best measured in two places:

1. In a straight line underneath the fingers (W1),
2. In a straight line from the skinfold just above the thumb to the outer edge of the palm (W2).

The length of the phalanxes of each finger and their proportions to each other are of special importance. They must, of course, be compared with the length of the whole hand. While studying the hands of schizophrenics and mental defectives, I found in a striking number of cases overlong finger-tips. This finding made me establish a finger-tip ratio, which is the measure of the length of the whole medius finger in proportion to the length of the finger-tip.

$$\text{FINGER-TIP RATIO} = \frac{\text{length of the whole medius finger}}{\text{length of finger-tip}}$$

The normal finger-tip ratio is between 3·8 and 4·0. The length of the fingers in general, and of the terminal phalanxes in particular, is related to the hand-type. It is the elementary regressive hand where one finds most often an abnormal finger index. Overlong finger-tips go with a schizoid make-up and they are, as I have already said, most frequent in schizophrenic patients.

As everybody knows the medius finger is always the longest in a human hand. Then follows either the index or the fourth finger, then the thumb and finally the fifth finger. In those cases where the index is second in length we speak of a human digital formula, the term being coined by Wood Jones; in cases where the fourth finger takes second place one speaks of a simian digital formula, as in apes and monkeys the fourth finger is always longer than the index.

The thumb and the fifth finger, which are also called the marginal fingers, have a special place: they are more variable in length and general shape than the others. A misshapen thumb, for example, which is either too small or has a clubbed terminal phalanx or both is of a special psychological significance. It generally occurs in the elementary regressive hand and is one of the well-known typical traits in the hands of mongolian imbeciles. In the latter one also finds always an abnormally short fifth finger, which is often crooked at the same time. These abnormalities appear to be stigmata of degeneration and so is the ankylosis of the fifth digit or of all fingers. Ankylosis of one or more fingers is frequently met with in mental defectives. The misshapen thumb also occurs frequently in mental defectives, but rarely in schizophrenics.

While the thumb is linked with intelligence, the fifth finger has a particular correlation to endocrine functioning. The fifth finger is too short and deformed, for example, in congenital hypothyroidism and in hypopituitarism (Froehlich syndrome). In the hands of eunuchs and eunochoid people it is too long.

The index finger, because of its evolutionary development in man, is linked with the intelligence and it is a matter of fact that the 'human digital formula' is very rarely found in a mental defective.

The length of the medius finger is the most stable one. It is, except for a few phenomenally rare cases, always the longest finger. There are a number of hands of schizophrenics, where the medius, the index and the fourth finger are of almost equal length. This anomaly is of important diagnostic significance.

THE PHYSICAL QUALITIES OF THE HAND

As Purkinje has said in a treatise mentioned earlier in this chapter, the physical qualities of the hand are of important psychological significance. One means by them: the habitual temperature, humidity or dryness, colour and flexibility.

All physical qualities of the hand are, of course, subject to outside and organic influences. If we expose our hands to cold or heat, this will naturally determine their temperature and their colour. If we have a nervous shock, or any deficiency of blood circulation that will have the same influence, it will alter temperature and colour of the hand. If we have rheumatism or gout that will influence the flexibility of our hands and if we have a fever or any infectious illness that will have an influence on the degree of humidity of the palms. All this is obvious to the medically trained person. Apart from the factors which I have just described, there is undoubtedly, under 'normal' conditions, a very strong psychological factor which is responsible for the physical qualities of the hand.

The psychological significance of these physical qualities is easy to understand through the following consideration: the autonomic nervous system transmits emotional reactions to the circulatory system, the endocrine system and the sweat glands. It translates, so to speak, psychical events into physical symptoms. The latter which are visible or tangible can therefore reveal the former. The sympathetic as well as the parasympathetic branches of the autonomic nervous system act either separately or at the same time on the physical qualities of the hand. Their function is mostly antagonistic but may, when a certain limit of tolerance has been surpassed, change into the opposite, which means that parasympathetic reactions may change into sympathetic ones at a certain moment or vice versa. The parasympathetic nerves are mainly connected with those emotional reactions which one experiences as disagreeable, while the sympathetic nerves react in the opposite direction. This division is however a very crude one and not altogether exact, as certain agreeable emotions go with parasympathetic reactions. The simplification of the matter will allow the lay reader to grasp the essential point better than a complete presentation of the very intricate functioning of the vegetative nervous system, which would also take me too far away from my subject.

Temperature and colour of the hand are interrelated. Pale hands or white hands, which are typical of the melancholic or chronically depressed, are generally cold to touch, while pink or reddish hands are warm as a rule. The peripheric circulation is stimulated through widening of the blood vessels under the influence of the sympathetic nerves, which happens in practically all agreeable emotions and feelings. When the mind is at perfect ease or elated, the hands are warm and of pink or reddish colour. Warm hands are more often dry than cold ones. But this is not always the case. Sweating of the hands is well known as a symptom of shyness and embarrasses many timid people, especially at the age of adolescence when this symptom is typical. It goes at any age with anguish and fear or abnormal timidity.

The flexibility of the hand deserves special attention. While the other physical qualities depend on the working of the autonomic nervous system and are fluctuating, the flexibility of the hand is a more or less stable quality. When the flexibility of the hand is not determined by factors of physical illness, it is a mark of certain psycho-physical or constitutional characteristics. It is first of all a symptom of postural habits of the hand, which as Wallon, among others, has pointed out, are characteristic of mental attitudes. One of the most well-known examples of the hyperflexibility of the hand is that of the mongolian imbecile. It is a symptom of the unusual flexibility of the joints, which goes with certain postures and special vocational gifts: namely for acrobatics and dancing. It also appears to be connected with certain emotional and even social attitudes, namely plasticity of feeling and adaptability. The hyperflexible hand is, of course, not only found in these charming and unfortunate mongolian imbeciles, but occurs often in the asthenic person with small sensitive hands. It is rare in all other hand-types. Stiffness of the joints is found in the elementary regressive hand as a rule as well as in the bony motor hand. It is typical of psychotic patients and most types of mental defectives, namely those with a schizoid temperament. One can assume that the psychological significance of stiffness of the hand shows a predominance of introvert tendencies and a strong inhibitory factor. The elementary simple and fleshy motor hands are not highly flexible because of their strong bony structure, but they are not rigid hands. Endocrine function also influences the flexibility of the hand, as I have mentioned in the chapter 'The Endocrine Factor'. It is necessary to differentiate the flexibility of the palm from that of the fingers because they are in some instances different. This is, for example, often the case with hypopituitary functioning, where one finds a rather stiff palm and hyperflexible fingers, especially finger-tips. As endocrine function touches so closely the emotional make-up, the degree of flexibility of the hand in these cases can be regarded as both an endocrine symptom and a psychological one.

Instead of calling this chapter 'The Method of Hand-Interpretation' I should have named it 'How to make use of the data of the hand for a method of hand interpretation'. This would have been a more precise but a rather heavy and clumsy headline, and I can justify the title I have given by the fact that the classification and interpretation of the hand as a whole, and in all its parts, implies the teaching of a method of how to use this material. I also felt that to be more dogmatic and to prescribe a special way as to analysing hands would have been a mistake, as it would easily induce the more sanguine and psychologically untrained reader to try 'his own hand' on the subject. That is just, as I have already pointed out, what I want to prevent. There must be no misunderstanding about the fallacy of the idea that any knowledge can be safely acquired from reading a book. The danger of such a dilettante attitude must be realized. In ending the chapter I describe nevertheless the general outline of the proceeding of a hand-interpretation as I do it myself. As I have safeguarded myself in the preceding phrases, the following indications may be useful to those physicians, psychiatrists and psychologists who know enough about the subject to try my method.

In every case, the careful inspection of the whole hand and the necessary measurements, as I have indicated them, should precede any detailed investigation of the papillary ridges and the creases. Notes on the physical qualities and on the nails, their shape and other qualities, are essential. The best way to investigate the hand at leisure is to take photographs of the backs and prints of the palmar surfaces.

It is of no essential importance in which order one proceeds with a hand-interpretation but I found the following method of doing it very useful:

1. GENERAL INSPECTION OF THE HAND (DORSAL AND PALMAR SURFACE).
2. MEASUREMENTS OF THE HANDS.
3. COMPARISON OF HAND MEASUREMENTS WITH HEIGHT AND WEIGHT.
4. CLASSIFICATION OF THE PRINCIPAL HAND-FACTORS OR STRUCTURAL ELEMENTS.
5. DIAGNOSIS OF THE ENDOCRINE FACTOR.
6. INVESTIGATION OF THE PHYSICAL QUALITIES OF THE HANDS.
7. INVESTIGATION OF EACH FINGER.
8. CAREFUL INSPECTION OF THE NAILS.
9. INSPECTION OF THE PALM:
   (*a*) THE MAIN CREASES,
   (*b*) THE ACCESSORY CREASES,
   (*c*) THE PAPILLARY RIDGES.

With these data a qualitative estimate can be made of the intelligence especially of the disposition towards a more concrete or a more abstract way of thinking; the imaginative factor can be assessed and also its influence on the intellectual qualities as well as on the emotions.

A general estimate can be given of the individual endocrine formula and of the temperament in its various aspects. From all these investigations one can arrive at a diagnosis of personality and of its capacity of integration.

As I have repeatedly said, there is not one psychological test which can tell us all that there is to be known about a person. Other tests must assist that of hand-interpretation in order to get a complete picture of the person concerned. But I hope that I have been able to make it sufficiently clear in the general part of this book that the hand is one of the most useful guides to the diagnosis of the constitutional tendencies of man.

CLINICAL STUDIES

CHAPTER VI

THE HAND OF THE MENTAL DEFECTIVE

A. PRELIMINARIES

THE term mental deficiency gives rise to a wrong or at least to an incomplete idea about the mental defective, to which not only the layman, but also the ordinary physician is subject. An appropriate idea of mental deficiency can only be got from the conception that it represents only one aspect of a general psycho-physical deficiency. One should therefore rather speak of a defective personality with subnormal intelligence than of a mental defective. Both diagnosis and therapy have suffered from this limited notion to the disadvantage of the patient as well as of the community; for the high-grade mental defective, if rightly treated and properly educated, can, under certain circumstances, be of service to others. Apart from common sense it seems to me to be inhuman to look at a mental defective as on an outcast because he is inferior in intelligence. He remains in spite of this deficiency—except when it removes him too far from human contacts—a person with a personality. It is important to know about this personality, and other tests than those of intelligence only, ought to be applied, adapted to the peculiarities of people with different degrees of mental inferiority. It is common knowledge that psycho-motor reactions of the moron and the imbecile differ from those of other people and could well be used for diagnostic purposes and for therapy on physical lines, namely through education in motor co-ordination and muscular relaxation in those cases where there is a chance of improvement, as in the feeble-minded and high-grade imbecile. It is evident that subnormal intelligence goes with inade-quate thinking as well as poverty of thought, and for this reason alone psycho-motor reactions are confined to a much smaller range than in normal people and exist mainly on the automatic level. The interchange between the mind and the milieu becomes more and more narrowed in proportion to the severity of the mental deficiency and ceases altogether in idiocy. I have tried to show in the first chapter of this book the intrinsic link which exists between thought and psycho-motor functions, especially those of the hand; this is at the same time an explanation of the psycho-motor deficiencies observed in mental defectives. Naturally a poor and defective intelligence is linked with a poor imagination and lack of foresight. The latter may also explain the bad co-ordination and impulsivity of movements. Intelligence and imagination are both linked with perceptive faculties, which are of course also faulty in the mental defective. The sense of touch, for example, which is essential for the acquisition of the knowledge of objects, cannot be used in the same way as in intelligent people, namely as a channel between the outer and the inner world on which the development of the mind depends. Habitual hand movements, which depend on perceptive qualities and on thought, are, therefore, restricted and lacking in co-ordination in the mental defective, and this must reflect on the muscular development of the hand as well as on its crease-lines. It is also well known that the mobility of the fingers in mental defectives is restricted and stereotyped, even in high-grade imbeciles and feeble-minded people, who have difficulty, for example, in moving one hand or one foot separately. Their movements are synkinetic, which means that a number of unnecessary, supporting gestures are used, as a very young child does whose sense of balance is not properly established. The movements are global and lack differentiation, especially those of the hand. There is however one exception to the rule. One finds in patients who suffer from idiocy with Bourneville disease the most extraordinary mobility of the fingers. These idiots move a little piece of cotton wool or paper between the thumb, index and medius fingers with an acrobatic facility, that no normal person could ever imitate. Dr. C. J. C. Earl has described this phenomenon in a paper 'The Primitive Catatonic Psychosis of Idiots'. In three cases I have taken prints of their hands, which are very narrow with long fingers and numerous accessory creases on the distal part of the palm, around the index and medius fingers. The compulsive movements which the patients perform for hours apparently without getting tired, can be understood as the result of a kinaesthetic impulse below the level of consciousness, resulting from a profound nervous tension which discharges itself in this strange mannerism. One can compare these gestures with the compulsive mannerisms of other types of psychotics, notably schizophrenics. The unusual number of half-circular lines on the distal part of the palm (see print 30) demonstrates that habitual movements of the hand are reflected on its crease-line pattern. This can take place even at a certain age, as it is reasonable to assume that the patients did not perform the described gestures when they were small children.

Speech which can be regarded as a specialized psycho-motor function is retarded and inhibited in the mental defective. Dr. Leo Kanner points out in *Behaviour Disorders in Childhood* how speech difficulties of this kind are associated with graphic difficulties. He says:

> It is a matter of common observation that inherent intellectual inadequacy (mental deficiency) is associated with corresponding difficulties in the acquisition of verbal tools for symbol formation and communication. Late onset of speech is followed by deficiencies both active and passive in vocabulary and in the ability to learn graphic symbolization through reading and writing. This difficulty is, roughly speaking, proportionate to the degree of the intellectual defect. Speech does not only begin at a later age than is usual but articulation, enunciation, grammatical construction, ease of expression and the capacity for the grasp and use of abstractions is impaired in varying degrees.

The restricted psycho-motor capacity of the mental defective determines his habitual postures and gestures in general and those of the hand in particular. The latter influence in their turn the development of the muscles—and the design of the crease-lines of the palm.

One has to distinguish three sets of hand-traits:

1. THOSE RELATED TO THE BONY STRUCTURE
2. THOSE RELATED TO THE MUSCLES
3. THOSE RELATED TO THE SKIN

1. *The Bony Structure of the Hand*

is prenatally conditioned and depends largely on genetic and endocrine factors.

2. *The Muscular Development of the Hand*

is up to a point interrelated with the shape of the hand, but depends very largely on the habitual movements and postures of the fingers. The crease-lines depend on both the movements of the hand and the direction of the joints. While every child is born with the thenar-line and the two transverse creases, the long longitudinal line is not necessarily congenital and the accessory creases develop and alter during lifetime. The latter show highly individual variability, due to the individuality of habitual movements and postures of the fingers, by which they are originated. I remind the reader of the four- or five-fold semicircular creases which arise in the hands of idiots with Bourneville disease as a consequence of habitual idiosyncratic movements of the first three fingers.

3. *The Third Set of Hand-Traits*

concerns

(*a*) The papillary ridges, the direction and spacing of which is entirely individual and never changes from birth to death.

(*b*) The nails, which consist of modified epidermis. The formation of the nails is conditioned by genetic and endocrine factors, but their shape, texture and colour can alter much during lifetime under the influence of physiological conditions. It is well known that the shape and colour of the nails alter in heart and pulmonary diseases and return to their original form after the illness has been overcome.

While the first and third set of hand-traits show the constitutional factor definitely and clearly, the second entails at the same time a marked psychogenetic factor. Both the muscles and creases of the palm convey certain traces of the interaction between the individual endowment and outside stimulation.

B. BACKGROUND OF THE STUDY

The study of the hand of the mental defective has occupied a very great part of my research work. I started it in 1934 in Paris, where I had the good fortune to attend regularly the consultations of Professor Henri Wallon for difficult and backward children. I visited at the same time regularly a Home for mentally defective children at Livry Gargan, where Professor Wallon was the supervising psychiatrist. During the last two years of my stay in Paris I also worked with Dr. Gilbert Robin and Dr. Mâle in their clinics for nervous and mentally defective children. Dr. Robin paid special attention to the endocrine factor in both groups of children and applied in many cases endocrine therapy, with encouraging results. My study in these clinics consisted of an analysis and diagnosis of the hands of the young patients in regard to their temperament and mentality. I saw the children before they went into the consulting-room of the psychiatrist. My diagnosis was thus independently made. My notes were compared with the statements of the psychiatrist and the results were very encouraging, as the percentage of agreement was between 75 and 85 per cent.

I realized fully for the first time during my studies with Dr. Robin the importance which the endocrine factor has in both nervous disturbances and mental deficiency. Through my work with Professor Wallon I learned about the correlation which exists between mental deficiency and psycho-motor deficiency.

When one starts to explore a new territory of research one is inclined to be somewhat carried away by the enthusiasm of discovery. I found however during my work in Paris a sufficiently safe basis of hand-interpretation for me to discover both the endocrine and the psychological factors from hand-features, but I concentrated more on the individuality and personality of the young patient than on clinical details. In time, especially after I had come to this country, my method of research shifted more and more from the individual to the clinical factor which is common to patients who suffer from the same type of nervous or mental disturbance. Soon after my arrival in England I continued my research on the hands of mental defectives at St. Lawrence's Hospital, Caterham, which I visited at intervals over a period of six years. During the last four years I went for several visits to a Colony for mentally defective children, to investigate especially the hands of high-grade boys and girls. I had there an opportunity to study the hands of the same children after intervals of one year, a study which is of particular interest in order to investigate the coefficient of growth as seen in the hand and also to observe if there are any changes in the pattern of the accessory crease-lines. I did not concentrate on the hands of children only, but saw also a great number of adult mental defectives at St. Lawrence's Hospital. The reader may ask why I have given so much time and paid so much attention to the hands of people who are of so little use to society. I have already pointed out in earlier writings that it is only through the study of the abnormal that we can hope to learn something about the normal personality. While I was in Paris my research was concentrated on the hands of children and adolescents, and with them it is of the greatest importance to discover the causes of emotional and social maladjustment, of which mental deficiency may be one. The problem of mental deficiency or rather that of the mental defective is both a medical and a social one and that seemed to me one more reason for studying this special subject.

When one thinks of mental defectives one has children in mind rather than adults, and when one walks through a big hospital for mental defectives, such as St. Lawrence's Hospital, one is so much more impressed by the children than by the grown-up patients. Their tragedy seems to be so much more poignant. Indeed, I confess that it took me quite a time to feel 'detached'. I was, to start with, very moved by the sight of the most severe cases. One sees there dwarfs walking through the wards slowly and badly, or completely under-developed bodies lying motionless with hands, legs and feet of small children. But no one must get the idea that Homes for mental defectives are places of horror. Not at all. The terrible tragedy is only present in a limited number of cases, who are confined to special wards and well separated from those who are able to have some kind of life. It was through the latter group that I realized how important it is to find a means of helping these lighter cases to some brightness and purpose of living. Quite apart from a study of their hands with a view towards a comparison with those of normally intelligent children, I saw the importance of procuring an accurate diagnosis of the degree of mental deficiency or, in border-line cases, to make sure if backwardness was really due to inferior intelligence. I made it my task to find out if such an objective assessment exists, by means of the hand. I have already mentioned in this book that there is always the possibility of error in assessing the I.Q. in feeble-minded persons, especially in children and therefore the danger that a normally intelligent child may be put into a Mental Home because it had failed in an intelligence- and performance test. When I visited the school at St. Lawrence's Hospital, I was struck by the eagerness of the children to learn and by their co-operative spirit as well as their great pride in good performances. This showed particularly in their handicrafts, for which there was a special class. Some of the results were remarkable, for example the needlework of the high-grade girls and the toys and baskets, &c., of the high-grade boys.

There was a separate house for feeble-minded adolescent boys who went out to work in the neighbourhood. The atmosphere, cleanliness and friendliness of this place made one feel that this was something approaching a happy home. The 'odium' of any Institution cannot, of course, be quite wiped out, and it must be the aim of the psychiatrist to get these cases fit to go out into the world and to lead a more or less normal life with a background that takes account of their special handicap.

The atmosphere of homeliness was even more accentuated in the colony for mentally defective children which I visited. It is a colony in beautiful grounds, with gardens and possibilities for sports and games. The children live in small houses, supervised by Sisters or male nurses, and have a great amount of freedom. In this colony there is practically nothing which one could call an 'institutional atmosphere'. Indeed, how much happier many children would be here than in the unhappy or sordid homes of asocial parents. The schooling is built on very modern ideas and the teaching is imaginative and individual. All children are tested in regard to their I.Q., but the psychologist takes the emotional factor, which so easily falsifies results, fully into account. The teachers have a real and personal contact with the children beyond and outside their lessons and certainly possess their confidence. The spirit of sympathy, authority and friendship between the Superintendent and the children has made a great family of this vast number of different young people, of different ages and with different mental and psychological defects. Never before had I realized, as I did here, how much this human and understanding attitude can contribute not only to the happiness but also to the psychological improvement of mentally defective children. It is not so long ago that many physicians and even psychiatrists discarded these patients as being rather a nuisance, whom one had better leave alone, as one cannot help them anyway. So they were left to wither away in the cellar atmosphere of lack of sympathy and of loneliness. One can safely say, that today such an attitude is the exception rather than the rule and the problem of the mental defective gets more and more the medical, psychological and social understanding that it needs and deserves. It is well known that many mentally defective children are of a very affectionate and placid disposition, especially the mongolian imbeciles, who are always the pets of the wards. There are some institutions where the mental defective is almost worshipped as a kind of superior being that is nearer to God than other mortals. I am speaking of the Homes of the Rudolf Steiner Group. While I think that their interpretation of mental deficiency as something 'superior' is tinged with sentimentality and is a misunderstanding of the real situation, I do appreciate tremendously the love and care that is given by the Rudolf Steiner teachers to these young and unfortunate people. We can be grateful to them, indeed.

THE STUDY ITSELF

There are few references in medical literature to the hand of the mental defective, in spite of the fact that hand-features in mongolian imbeciles are recognized as being marks of the illness. There is no book on endocrine glands which does not mention the 'mongolian' hand and yet the obvious correlation between a psychophysical condition, as mongolism is, and the hand has not instigated research on the hand in other mental disturbances on a large scale. Apart from mongolism the hand represents in cretinism one of the most significant symptoms of that condition. Here the peculiar shape of the palm and fingers is due to hypothyroidism and the same probably applies, according to Benda (1936), to mongolism, though S. L. Simpson does not share his view. In addition to these examples, gross deformities of the hand often found in mental deficiency are also well known and have been mentioned among other authors by Penrose (1931) and Rolleston (1936). The latter refers to the frequent occurrence of polydactilism in a special type of adiposo-genital dystrophy—the Laurence Biedl syndrome which goes with mental deficiency. R. M. Stuart (1939) described a case of arachnodactily in a mental defective and C. J. C. Earl has noted, as I have already mentioned, the hands of psychotic idiots with Bourneville disease (1934).

The shape of the nails in mental deficiency has only been treated by Pardo Castello in his excellent book *The Diseases of the Nail* (1936) and by myself. I have investigated the nails of mongolian imbeciles (1942) and of the mental defective in general (1944).

Very scanty are the references to the crease-lines and the papillary ridges in mental defectives. The dermatoglyphics of the palm and fingers in mental defectives have been mentioned by Féré, who pointed out that the finger-prints of degenerates show more dissimilarity than those of normal people and who noted the higher frequency of a primary form of papillary ridges on the finger-tips of epileptic imbeciles. H. Cummins (1939), who is well known for his research on the dermatoglyphics of monkeys and apes, found, in the hands of sixty mongolian imbeciles whom he investigated, unusual patterns of the papillary ridges on both palm and finger-tips.

Even less attention has been paid by scientists to the crease-lines of the hand, except for the so-called simian line in mongolian imbeciles. Of the numerous references to this peculiarity, I only mention the provocative book by Crookshank *The Mongol in our Midst* (1931).

E. Wilson is one of the very few people who have investigated the creases in the hands of mentally abnormal persons. She compared the crease-lines of the hands of forty-seven mental defectives with those of fifty schizophrenics and fifty College students of the

Columbia University, New York (1931). She found that the position of the lower transverse crease was normal in only 7 cases of mental deficiency in proportion to 30 cases out of 50 College students.

I have described the crease-lines of mental defectives in a study done on the hands of 512 cases from St. Lawrence's Hospital, Caterham, in a paper 'The Hand of the Mental Defective' (1944) and of 115 high-grade boys in a paper 'The Dermatoglyphics of the Hands of 115 High-Grade and Difficult Boys' (1947), and the reader will find some general information on the subject in my book *The Human Hand*. I make considerable use of both papers in this chapter, as they have complete bearing on the problem I am dealing with.

THE METHOD OF STUDY

I have taken hand-prints of all patients investigated according to a method described in earlier writings which I am repeating here in an abbreviated form. The hands are greased with a small amount of Nivea cream or any other face cream and pressed on a smooth paper placed on a rubber pad. The latter is put on a stiff piece of material upon which I have fixed a very small rubber pad corresponding to the hollow of the palm. The grease prints thus obtained are made visible by spreading a black powder (copper oxide) over them, which reproduces the whole system of creases and papillary ridges on palm and fingers. The prints are then fixed with the fixative used for charcoal drawings.

The record of the palmar surface is, of course, insufficient and had to be complemented by notes on the back of the hand, the fingers and nails with all their peculiarities, and on the physical qualities. Measurements of the hands were taken according to the method I have described.

The material of study was then classified according to age, sex and the degree of mental deficiency. One distinguishes among mental defectives the feeble-minded, the high-grade defective, the medium-grade defective, the low-grade defective and the idiot.

The list of the hand-features as represented by the tables below, as well as their sequence, indicates the purpose of my investigation: an all-round study of the hand of the mental defective from its structure to its physical qualities. The sequence of the hand-traits listed is in direct correlation to their significance for mental deficiency. While the structure of the hand reveals much about the working of the endocrine system, which is an especially important factor in mental deficiency, the functioning of the autonomic nervous system, as diagnosed from certain physical qualities, shows specific and habitual emotional reactions of the patients. The shape and quality of the nails reveal a lot about genetic and health factors and have a definite significance in mental deficiency. The muscular development and the creases are characteristic of psycho-motor habits, and as these are abnormal in the mental defective, their shape and form is indicative of both: restricted emotional impulses and restricted emotional expression. As the shape of the fingers is very characteristic of intelligence, the study of the digits has taken a predominant place in this investigation. The papillary ridges show often special forms in the mental defective and have therefore also a diagnostic significance. These considerations may suffice to emphasize the value of hand-interpretation in diagnosing mental deficiency. As many of the hand-traits enumerated in the tables are already established at birth, the diagnosis of mental deficiency can be made at a very early age, practically after birth. In any case, a child who does not seem to develop normally can be easily tested by means of his hands and the diagnosis may then be verified by other tests.

The view that hand-diagnosis is of a very great diagnostic value has been stressed by some scientists of authority, such as Wood Jones, Osseretsky and Noiré. Noiré, for example, called the hand the visible part of the brain. Osseretsky stressed the fact that the whole body in its static as well as its dynamic functions adapts itself to the functions of the hand and states that the study of the hand is therefore of the greatest importance to psychology.

THE CATERHAM CASES

The greater part of the material which I have collected on the hands of mental defectives comes from St. Lawrence's Hospital, Caterham, and I therefore refer to all data connected with my work there as the data of the Caterham cases.

THE MATERIAL CONSISTS OF OBSERVATIONS ON:

122 FEEBLE-MINDED PATIENTS
347 IMBECILES WITHOUT MONGOLISM
43 IDIOTS.

Among the 347 imbeciles were:

31 HIGH-GRADE IMBECILES
129 MEDIUM-GRADE IMBECILES
187 LOW-GRADE IMBECILES.

There were more female than male patients; the distribution according to sex being:

	FEEBLE-MINDED	IMBECILES	IDIOTS
male	44	102	20
female	78	245	23

The distribution according to age was:

AGE	IN 122 FEEBLE-MINDED PATIENTS	IN 43 IDIOTS
5–10 years	5	3
10–16 ,,	25	8
16–21 ,,	18	7
21–30 ,,	36	17
30–40 ,,	25	7
over 40 ,,	13	1
	122	43

AGE	IN 347 IMBECILES		
	HIGH-GRADE	MEDIUM-GRADE	LOW-GRADE
5–10 years	3	4	9
10–16 ,,	6	28	38
16–21 ,,	3	19	14
21–30 ,,	7	28	56
30–40 ,,	5	17	30
over 40 ,,	7	33	40
	31	129	187

I have taken 150 normal people as control cases for the statistics of Table I, while in the original paper, 'The Hand of the Mental Defective', I only had 100 hands of normal people compared with those of the mental defectives. It was unfortunately not possible to find equal numbers of cases for each group of mental deficiency, which would have been the ideal situation, but the data relate to a sufficient amount of material to be significant, especially in view of the additional research done at the 'Colony'.

I paid special attention during my investigations to those hand-traits which are already well known and listed them together with those I discovered through experience as being of equal or greater significance.

Table I deals with 15 hand-traits which are mostly typical of mental deficiency. Tables II and III deal with the crease-line system and its peculiarities in mental deficiency, while Tables IV and V are concerned with the physical qualities.

TABLE I

Percentage Incidence of Certain Hand-Features in Mental Defectives and in 150 Control Cases

	122 FEEBLE-MINDED	160 HIGH- + MED.-GRADE	187 LOW GRADES	43 IDIOTS	150 CONTROL CASES
1. Too small thumb	19·7	25·9	38·3	51·1	3·0
2. Simian-like insertion of thumb	1·6	1·8	2·6	4·6	0
3. Too short fifth finger	13·1	24·7	32·8	41·6	10·0
4. Ankylosis fifth finger	3·3	3·8	11·1	11·6	0
5. Lack of one crease on fifth finger	0·0	1·2	2·8	9·1	0
6. Abnormal digital form	1·6	3·9	5·8	11·6	0
7. Syndactylism	0·0	0·0	0·52	9·06	0
8. Ankylosis of all fingers	0·0	0·63	2·6	4·4	0
9. Rudimentary nails	20·4	27·8	43·9	44·8	2·0
10. Concave primary nails	5·7	5·7	15·3	12·2	0
11. Arc-pattern pap. ridges on thumb	4·9	3·1	11·6	20·4	6·0
12. Arc-pattern on finger-tips	26·2	13·7	20·8	46·04	20·0
13. Whorls or loops on thenar or hypo-thenar eminence	22·9	24·6	29·6	19·0	15·0
14. Simian line	9·8	34·4	15·9	11·6	2·0
15. Simian-like line	29·9	29·07	31·2	22·4	13·0

TABLE II

Percentage Incidence of Abnormality of Crease-Lines in Mental Defectives and in 150 Control Cases. The Data in this Table refer to

	15 FEEBLE-MINDED BOYS	43 OTHER FEEBLE-MINDED PERSONS	157 HIGH- + MED.-GRADE IMBECILES	50 LOW-GRADE IMBECILES + IDIOTS	150 CONTROL CASES
1. Long longitudinal line lacking	31·2	18·6	22·9	46·0	5·0
2. Abnormal direction of one main crease other than simian line	37·4	65·1	43·1	68·0	5·0
3. Confused crease-line pattern	0·0	0·0	3·1	31·2	0·0

TABLE III

Percentage Incidence of Defects in the Two Main Transverse Creases. The cases studied are the same as in Table II

	15 FEEBLE-MINDED BOYS	43 OTHER FEEBLE-MINDED PERSONS	157 HIGH- + MED.-GRADE IMBECILES	50 LOW-GRADE IMBECILES + IDIOTS	150 CONTROL CASES
1. Ruptures in lower transverse line	31·2	60·9	45·9	84·0	19·0
2. Ruptures in upper transverse line	18·7	30·2	29·1	48·0	27·0
3. 'Isles'* in lower transverse line	31·2	44·1	39·1	74·0	22·0
4. 'Isles'* in upper transverse line	5·5	11·6	26·1	30·0	20·0
5. Lower transverse crease too short	12·5	6·9	5·1	25·0	1·0
6. Upper transverse crease too short	0	2·3	3·1	10·0	0
7. Other defects of lower transverse line	5·5	9·3	12·8	20·0	4·0
8. Other defects of upper transverse line	5·5	6·9	4·5	16·0	0

* *Note:* one understands by isle a circumscribed, more or less oval space in a crease-line.

Tables IV and V show the correlations between the physical qualities of the hand and mental deficiency.

TABLE IV

Percentage Incidence of Abnormal Colour, Cold and Moist Hands in 116 Imbeciles of different age-groups

AGE	ABNORMAL COLOUR	COLD HANDS	MOIST HANDS
5–15 years	28·2	68·7	65·6
15–25 ,,	44·0	48·0	33·0
25–35 ,,	30·6	47·3	47·4
over 35 ,,	28·6	31·5	31·5

TABLE V

Percentage Incidence of the Degree of Flexibility of the Hand in 81 Feeble-minded, 251 Imbeciles and 20 Idiots

1. STIFF HANDS

AGE	81 FEEBLE-MINDED	251 IMBECILES	20 IDIOTS
5–15 years	45·8	42·3	63·6
15–35 ,,	60·0	58·0	71·4
over 35 ,,	78·2	52·3	71·4

2. HYPERFLEXIBLE HANDS

AGE	81 FEEBLE-MINDED	251 IMBECILES	20 IDIOTS
5–15 years	20·8	20·2	18·1
15–35 ,,	2·9	9·6	0·0
over 35 ,,	0·0	9·2	0·0

DISCUSSION OF THE TABLES OF THE CATERHAM CASES

TABLE I

Among the fifteen items in this table, ten represent abnormalities of the fingers and nails and only five concern the papillary ridges and crease-lines. One glance at the percentage incidence of hand-traits in the first group shows their significance for the diagnosis of mental deficiency. Syndactilism is of very rare occurrence in normal people and is most significant for mental deficiency and so is the ankylosis of all fingers. When the latter only concerns the fifth finger *one must be careful to come* to diagnostic conclusions. This hand-trait is frequent in psychotic patients, particularly schizophrenics, but I have also seen it in people without either psychosis or mental deficiency. Ankylosis of the fifth finger has a genetic significance as it is an inherited trait. An abnormal digital formula which is relevant for the diagnosis of mental deficiency according to this table, also appears in the hands of schizophrenic patients, mostly in primary psychosis with mental deficiency, and I refer to this hand-feature more fully in the chapter 'The Hand in Mental Illness'.

The simianlike insertion of the thumb, which means that the thumb is placed abnormally low in the hand, is of a 100 per cent relevance. The condition is like that in monkeys and apes and as with them the first phalanx of the thumb is much shorter than is usual in normal human beings. Print 25 which represents the palmar surface of a female idiot illustrates exactly the position described. This abnormality has, to my knowledge, never been described before, while the other hand-traits so far mentioned are all well known to medical men.

The lack of one crease-line on the minimal digit, which Penrose has mentioned in correlation with mongolian imbecility, is highly significant for mental deficiency, and not only for mongolism, as this table demonstrates. This special hand-feature never occurs in psychosis or in normal people.

Concave, primary nails are most significant too.

The abnormally short thumb and fifth finger, so frequently present in mental defectives, occasionally occur in people with normal intelligence, the explanation being that both traits are connected with endocrine dysfunctions. One finds the too short fifth finger in hypothyroidism, which can, but must not be associated with mental deficiency.

The last five items of Table I represent abnormalities of the papillary ridges and crease-lines.

These are much more frequent in mental defectives than in normal people, but the difference is only one of degree and these traits are not by themselves of diagnostic value, but represent a supporting factor for diagnosis when other more significant traits are present. The simian line has already been discussed in another context in this book. Its frequency increases with the degree of mental deficiency, but it appears also in highly intelligent people with neurotic tendencies.

DISCUSSION OF TABLE II

The item of this table, which is most significant for mental deficiency, is what I have called a 'confused crease-line pattern'. It is only present in the hands of low-grade imbeciles and idiots and mostly in mongolian imbeciles. The term chosen calls for an explanation. A confused crease-line pattern is distinguished from a normal one through the fact that there is no appreciable distinction between the main and the accessory crease-lines, by which the whole palm is covered. They are all of the same design and lack any accentuation. There is also not the usual preference of certain parts of the palm for quantity and intensity of creases. This lack of any differentiation makes the palmar surface appear like a drawing by a very small child. The lines are clumsily 'designed' and hesitantly drawn.

The absence of the long longitudinal line is significantly higher in mental defectives than in people of normal intelligence, but this hand-trait is not sufficient in itself to represent a condition of mental deficiency. One finds it also in highly intelligent people, in whom it occurs more often than the simian line. I have already pointed out that both these hand-traits are generally observed in the elementary regressive type of hand and belong to the same syndrome of hand-features.

The abnormal direction of one of the main crease-lines is also only relatively significant for mental deficiency.

DISCUSSION OF TABLE III

I was interested to check the findings of E. Wilson about the frequency of certain faults in the main creases. Her results have been confirmed by this investigation. Such defects as ruptures and isles are far more frequent in the pathological than in the control cases and appear mostly in the hands of idiots. They can therefore be regarded as significant for mental deficiency, but as they also occur in the hands of normally intelligent people, they cannot be used by themselves as a basis for diagnosis. This restriction holds for most single hand-traits correlated with mental deficiency. There are, however, a very few which by themselves are of absolute significance.

The whole of the hand has always to be taken into account and the value of the single traits must be considered and balanced with that of all other hand-features.

It is interesting to note that too short a lower transverse crease-line is rarely found in persons of normal or high intelligence.

Among 'other defects', I have listed such peculiarities as an abnormal ending of a main crease-line, for example its broadening at the end instead of thinning.

In mental defectives who are profoundly disintegrated personalities, one can expect to find defects in the flexor lines of the hand for the following reason: transverse creases are due to movements of flexion. The predominance of the latter over movements of extension corresponds, as Goldstein maintains, to well-integrated or centralized personalities. If, as in the mental defective, movements of the hands are halted, clumsy and incomplete and extensor movements constantly interfering, so to speak, with flexor movements, abrupt and faulty transverse creases will be the natural outcome. The predominance of flexor movements represents not only a well-controlled motor activity but applies at the same time to a definite type of nervous and emotional response. A person in whom flexor movements of the hands prevail, possesses a certain stability of mood and a genial temperament as well as a well-balanced central and autonomic nervous system. On the other hand, emotional disturbance is expressed predominantly by extensor movements and incomplete and jerky gestures. This condition is found in the 'introvert' personality with a very high factor of emotional inhibition. The hand-forms characteristic of extreme introverted tendencies are the sensitive long and the motor bony hands, I have reproduced the hand-print of a neurotic 'introverted' person of a leptosome physique (see print 13). This man was suffering from a severe anxiety neurosis. All his movements and particularly those of the hands, were jerky and he told me that he had suffered from bad motor co-ordination and arhythmic movements since the age of 4 years. He also developed a stammer at the same age. Note the uncertain design of the crease-lines. They are too few in number and too superficially drawn and the lower transverse line is too short. He is a very disintegrated personality, backward in learning but of normal intelligence.

One can say that in every individual where there is a lack of integration with predominantly subcortical expressions, as in the mental defective and certain neurotics, defective principal creases and also accessory ones are the natural consequence.

DISCUSSION OF TABLES IV AND V

The items of Table IV, colour, temperature and humidity, closely depend on the working of the autonomic nervous system and the blood circulation. It is a well-known fact that emotional as well as mental disturbances directly influence the autonomic nervous system. The pale and cold hand of the depressed and the melancholic person, is, as I have already mentioned earlier, of common occurrence and demonstrates the constrictory influence of the vasomotor nerves. The reddish colour of the hands of many schizophrenics has been recognized for a long time as a typical vasomotor symptom of the illness and the moist hands of many normal children and adolescents is familiar also to the layman. It shows high excitability of the autonomic nervous system with a special effect on the sweat glands. In the mentally defective patients investigated one finds the high percentage of 68·7 per cent of cold and 65·6 per cent of wet hands in young imbeciles between 5 and 15 years of age. Abnormal colouring was found in 28·2 per cent, which is very high. In the same group, the highest percentage of abnormal colour occurred in juvenile imbeciles between 15 and 25 years of age. Cold and moist hands were most frequent in the youngest group. These results indicate a marked vasomotor instability in mental deficiency.

TABLE V

In regard to the flexibility of the hand, the most striking result is the high percentage of stiff hands in the young mental defective. The flexibility of the hand normally decreases with age, and its stiffness is therefore a normal feature in the older mental defective. The high percentage of stiff hands in the young patients might be explained as an unusual state of nervous tension conditioned by a superimposed psychosis or in a great number of cases by an epileptic condition.

SUMMARY AND CONCLUSIONS

I reproduce here the summary and conclusions as they are set out in my paper 'The Hand of the Mental Defective', except for the passages which deal with endocrine hand-traits, because I have largely revised my ideas on this subject through my research work with Dr. S. L. Simpson.

To start with, I summarize briefly certain conclusions which show definite correlations existing between mental deficiency and the hand.

I. (*a*) THE STRUCTURE OF THE HAND (TABLE I)

Ankylosis of the fingers; simianlike insertion of the thumb; lack of one crease on the fifth finger; severe abnormality of the digital formula; syndactilism; concave, primary nails.

These traits appeared to a greater or lesser degree in the hands of mental defectives of all groups but were completely absent in the hands of the control cases.

(*b*) THE CREASE-LINES OF THE HAND (TABLES II AND III)

The confused crease-line pattern is peculiar to the extreme of mental deficiency.

II. HEREDITARY TRAITS

Although I have studied the genetic factor only little in this book, outstanding abnormal hand-features in shape, nails, crease-lines and papillary ridges are apparently hereditary. It will be remembered in this connexion that the first proved case of Mendelian inheritance in man was that of brachydactyly (middle joint absent in each finger). The genetic factor must not be overlooked in diagnosis from the hand any more than it can be ignored in purely medical examination.

It should be sufficiently evident that confirmation of mental deficiency is found to a marked extent in the hand of the mentally defective person. Indeed, in case of gross deficiency, this is apparent to the untrained eye. But it should be emphasized that to the scientific observer gross and less-gross abnormalities of the hand as a whole are also indicative of the degree of mental deficiency.

It will be apparent, therefore, from this latter statement, that diagnosis by the hands in border-line cases can be of great use. For, as it has already been pointed out, the I.Q. test cannot be regarded as absolute in such cases. The nature of the test itself as well as the conditions of the testing situation can be largely impaired by inhibitory mechanisms in the subject which tend to invalidate the results. This is more particularly true of those cases where there is a neurotic constitution or emotional instability, either or both of which may well simulate mental deficiency.

THE 'COLONY' CASES

After I had finished my research work at St. Lawrence's Hospital, Caterham, I left for a while the study of the hands of mental defectives and investigated those of psychotic patients of all types at St. Bernard's Hospital, Southall. I saw there, of course, a number of mentally defective patients too, as psychosis and epilepsy is often superimposed on mental deficiency. This study helped me to differentiate certain hand-features characteristic of psychosis from those belonging only to mental deficiency. After two years of research at St. Bernard's Hospital, I returned to my study of the hands of mental defectives with new ideas and a better plan of carrying it out. I went to a colony for mentally defective children for several visits and investigated there the hands of 175 cases, 115 boys and 60 girls, all of them high-grade defectives, many of them with delinquent tendencies. This study was more specialized than the previous one and has more direct and practical relation to social problems, as most of the boys and girls concerned leave the Colony at the age of 17 years and are settled into a more or less ordinary life if they are fit for it.

THE STUDY OF THE 'COLONY' CASES

Apart from the 175 children who were either feeble-minded or high-grade imbeciles, I saw another 100 cases of all degrees of mental deficiency as well at the Colony. I enlarged thus the material and experience I had gained on the subject in my earlier studies. Both the psychiatrist and the ordinary person are aware that at the present day the problem of juvenile delinquency is a world problem. It is far beyond the scope of this book to discuss the sociological and psychological aspects of this tremendous question; nor is it possible even to begin to quote the great amount of literature available on the subject. The study of juvenile delinquency has its special place in psychology and there are a number of Institutions devoted to research on this subject as well as to the care of the young offender. During the last ten years or so psychology has made a hesitant and slow but an unmistakable entrance into the Juvenile Courts and consequently the psychiatric explanation of the crime and of the delinquent's own psychological difficulties is more frequently taken into account. In this way the psychologist has gradually begun to assist legal administration in spite of the prejudice against him shown by many, both judges and others. It is of the utmost importance to consider carefully how far juvenile delinquency is due to environmental—and how far to constitutional—factors. I think while both these factors undoubtedly play an important part, it is not possible to assess exactly the share of each. I am inclined to give prevalence to the constitutional factor, but this might be a professional bias on my part, as my work is concerned with constitutional psychology. Among constitutional factors mental deficiency has been generally accepted as one of the contributing causes to juvenile delinquency. It is of course both the mental and the emotional deficiency and the lack of foresight of these high-grade defective children, which are the real cause for their irresponsible actions. This could be seen most clearly in those boys at the 'Colony' who could not resist stealing or breaking out of the Institution, who were compelled to burgle, to commit arson, &c. The psychiatrist who treats a number of these boys psychologically is of the opinion that their failings are the outcome of their lack of foresight and of the compulsive reactions found in markedly disintegrated personalities. This is especially the case in the schizoid type of person who was by far the most frequent in the group of high-grade boys studied. I had the opportunity to verify this interpretation of the psychiatrist through a personal study of their behaviour, gestures and facial expressions at 'social' gatherings. A group of boys would come every afternoon to the house of the medical Superintendent like any other ordinary guests or friends. The boys were sitting round the fire, some five or six of them, some smoking cigarettes and all talking with complete freedom. They were so much at ease that they did not hesitate to tease each other or even the 'Doctor' and myself and there was a spirit of community and enjoyment of the party, that brought out their personalities to the full. Many of them had real charm and knew how to use it, others were shy and retiring. In this way one met with the same emotional atmosphere as in any other group of young people. What made this group different from a 'normal' one was however the unrestrained spontaneity of expression, such as much younger children might show, and also certain abnormalities of gesture and of facial expression. I noted, for example, an abundant use of gesture by most of the boys, especially by the more excitable ones, and that these were jerky, over-expressive and badly co-ordinated. Facial expression showed mental strain; twitching of the mouth occurred in a number of cases and movements of the forehead gave their young faces an especially tortured look. Synkinetic movements of hands and feet accompanied by facial movements were the rule, as I have already mentioned earlier in this chapter. This description may suffice to give a general outline of the group-personality of these high-grade defective children and to show the effect of both emotional and mental deficiency on their expressive behaviour. The compulsive character of the former and the disintegrating effect of the latter give sufficient explanation for delinquent tendencies on constitutional grounds. The 'high-grade' boy is lacking in foresight and judgement and, because of this, mental deficiency entails moral deficiency. Also the compulsive character of emotional reactions necessarily goes with lack of control in any situation that arouses emotion or excitement. These considerations are sufficient to explain on constitutional grounds alone the delinquent tendencies. Naturally, environmental conditions cannot be overlooked and the family histories of many of these boys provide in themselves enough proof of the devastating effect of insecure and asocial home conditions.

My realization that the primary reason for abnormal behaviour was the emotional make-up was what induced me in this whole study to lay special stress upon the correlation between hand and temperament. But all the investigations and observations on the hands of the boys I have just described were made by me when I knew nothing of them but their names and ages, and my diagnoses of their personalities were made before I was brought into personal contact with them.

HISTORY

There is no literature whatsoever on the correlation between the hand and personality in high-grade defective boys with delinquent tendencies.

THEORETICAL REMINDERS

In order to show the theoretical structure on which this study is made I must briefly repeat some theoretical considerations, which are directly relevant to this investigation.

1. ABOUT THE CORRELATIONS BETWEEN THE HAND AND ENDOCRINE GLANDS

The pituitary gland and the gonads determine in the first place the structure and the size of the hand. Eunochoid development, for example, goes with a late closure of the epiphyses which produces a long and narrow palm with long fingers. In certain forms of hypopituitarism, as for example in the Froehlich syndrome, the closure of the epiphyses is early and the hands are short and have short fingers with tapered or pointed finger-tips. The fifth finger is always pointed and very short. The hand remains also small in one form of hypopituitary infantilism and corresponds to what I have called the small sensitive type of hand. The astonishing fact, however, exists that there is quite a different form of hand correlated with a different form of pituitary infantilism, namely a rather long hand, with a broad lower and a narrow distal part of the palm and long fingers, which is a special variety of the elementary regressive hand. This type of hand goes with sexual and general infantilism, and was found in 44 out of 115 of the subjects studied and was by far the most frequent hand-type of all (see Table I).

2. ABOUT THE CORRELATION BETWEEN THE MUSCULAR DEVELOPMENT OF THE HAND AND HABITUAL PSYCHO-MOTOR REACTIONS

I have written at length about these correlations in several parts of this book and must only repeat here that both the muscles of the hand and the crease-lines depend on hand movements which are either movements of purpose (cortical) or movements of expression (subcortical). Broad and short hands are found in people in whom cortical impulses outweigh subcortical impulses. Long and narrow hands are found in those where the position is reversed. It is interesting to note that a great many hands of the boys investigated had well-developed thenar and hypothenar eminences, but the middle and distal parts of the palm were flat and the fingers were thin. There were very few accessory lines in this type of hand. These characteristics can be explained by the following: in the owners of such hands, movements of purpose are restricted to a very simple pattern and movements of expression are either very limited or confined to impulsive reflex movements representing the primitive level of expressive conduct. Such a state of affairs is characteristic of the primitive schizoid personality which was frequent among these high-grade difficult boys.

3. ABOUT THE CORRELATION BETWEEN THE FINGERS AND INTELLIGENCE

The reader will find in the first chapter of this book a theoretical explanation of the correlation of the fingers, and particularly the thumb and index finger are correlated with intelligence. I have also drawn attention to the fact that the physiognomy of the fingers, their mobility and the shape of the nails show definite and characteristic marks in mental deficiency. Penrose has observed that the crease-line of the distal joint of the fifth digit is frequently lacking in mongolism, and I have shown that the same condition occurs in mental defectives without mongolism. Because of these observations and the theoretical considerations described, the fingers have been studied extensively in the present investigation.

THE MATERIAL OF THE STUDY

The material of this study consisted of 115 boys, mostly high-grade defectives of the Colony. Their ages ranged from 9 to 16 years:

10	boys	being	between	9	and	12	years	old,
40	,,	,,	,,	12	,,	14	,,	,,
65	,,	,,	,,	14	,,	16	,,	,,

THE METHOD OF INVESTIGATION

Hand impressions were taken from the hands of all of the boys by the method described. The shape of the fingers and nails and the physical qualities were described in detail in every case. Measurements were taken of: (1) length of the hand, (2) width of the distal part of the palm beneath the fingers (W.1), and (3) width of the broadest part of the palm (W.2), (4) length of the single fingers, and finally (5) length of the finger-tips. The height and weight of the boys were noted in every case.

THE METHOD OF HAND-INTERPRETATION

The method of hand-interpretation followed a special design and was made in three stages:

1. *Inspection and Interpretation* of the structural features and physical qualities of the hand with the aim of assessing endocrine functioning.

2. *Inspection and Interpretation* of the muscular formation of the palm, the crease-lines and the papillary ridges on both palm and finger-tips with the aim of assessing habitual emotional reactions and temperament.

3. *Inspection and Interpretation* of the fingers with special regard to the finger-tips with the aim of diagnosing the presence or absence of mental deficiency.

Based on this method diagnostic statements were made:

(1) of endocrine tendencies and personality types; (2) of the intellectual capacity, and were compared with the psychiatric diagnosis of the Medical Superintendent of the Colony.

The sequence of the 7 tables which follow is arranged in accordance with the complexity of hand-features, starting with the most general characteristics of the whole hand and ending with the most detailed characteristics of the finger-tips.

TABLE I

Numerical Incidence of Certain Hand-Types in High-grade Defectives

ELEMENTARY REGRESSIVE LONG FINGERS	ELEMENTARY REGRESSIVE SHORT FINGERS	ELEMENTARY SIMPLE	LEPTOSOME	SENSITIVE SMALL	SENSITIVE LONG	GROSSLY DYSPLASTIC	TOTAL
44	26	5	12	9	4	15	115

TABLE II

Numerical Incidence of Different Finger-Types in Relation to Hand-Types

HAND-TYPE	*FORM OF THE FINGERS*			
	SPATULATE	SQUARE	TAPERED	MIXED
ELEMENTARY REGRESSIVE WITH LONG FINGERS (44 CASES)	27	8	5	4
ELEMENTARY REGRESSIVE WITH SHORT FINGERS (26 CASES)	16	4	5	1
ELEMENTARY SIMPLE (5 CASES)	1	3	0	1
LEPTOSOME (12 CASES)	8	2	2	0
SENSITIVE SMALL (9 CASES)	3	4	2	0
SENSITIVE LONG (4 CASES)	0	2	2	0
GROSSLY DYSPLASTIC (15 CASES)	2	3	10	0

DISCUSSION OF RESULTS OF TABLES I AND II

Seventy out of 115 boys had elementary regressive hands among which those with long fingers predominated. I have described the particular characteristics of the latter hand-type in this book and have mentioned that the owner of such a hand has a schizothyme personality of the dull variety. His emotional response is both obsessional and weak. The elementary regressive hand with short fingers was always found in boys with an athletoid physique, of a mixed personality type, but with predominantly extravert tendencies. Their emotional response was lively and their suggestibility was remarkably high.

Grossly dysplastic hands came third in frequency. Most of these hands were cases of acromicry and 6 were abnormally long. Three of the latter were eunochoid hands, and another 3 belonged to boys with hyperpituitarism. The acromicric hand goes as Friedemann (1928), Kretschmer and Wolff (1942) have shown, with a schizoid make-up and psycho-physical infantilism. The boys with eunochoid hands showed very weak instinctive drive and marked effemination. All three were constitutional homosexuals with paranoiac trends.

TABLE III

Numerical Incidence of Different Endocrine Hand-Types (Total 71 Cases)

	HYPOPITUITARY INFANTILE	MONGOLOID	TYPICAL MONGOLIAN	HYPOPITUITARY OBESITY	HYPOPITUITARY + HYPOTHYROIC	HYPOTHYROIC	HYPERTHYROID	EUNOCHOID
	32	15	9	3	3	3	3	3

Numerical Incidence of Certain Individual Hand-Traits found in Endocrine Hand-Types

	HYPOPITUITARY INFANTILE	MONGOLOID	TYPICAL MONGOLIAN	HYPOPITUITARY OBESITY	HYPOPITUITARY + HYPOTHYROIC	HYPOTHYROIC	HYPERTHYROID	EUNOCHOID
Abnormally short fingers	11	10	9	2	3	3	0	1
Thumb too small	18	13	9	1	2	0	0	1
Thumb too long	3	0	0	0	0	0	3	1
Fifth finger too small	17	13	8	3	3	3	0	1
Crease missing on fifth digit	0	1	0	0	0	0	0	0
Hyperflexibility:								
Palm	15	10	9	1	2	2	0	2
Fingers	20	12	9	3	2	1	0	2
Medium flexibility:								
Palm	10	3	0	0	0	0	1	1
Fingers	5	2	0	0	0	2	1	1
Stiffness:								
Palm	7	2	0	2	1	1	2	0
Fingers	7	1	0	0	1	0	2	0

TABLE IV

Numerical Incidence of the Degree of Flexibility of the Palm and Fingers in Relation to Hand-Types

HAND-TYPE	PALM			FINGERS		
	HYPER-FLEXIBLE	MEDIUM	STIFF	HYPER-FLEXIBLE	MEDIUM	STIFF
ELEMENTARY REGRESSIVE LONG FINGERS (44 CASES)	8	15	21	13	13	18
ELEMENTARY REGRESSIVE SHORT FINGERS (26 CASES)	15	6	5	21	4	1
ELEMENTARY SIMPLE (5 CASES)	0	3	2	0	3	2
LEPTOSOME (12 CASES)	1	5	6	0	5	7
SENSITIVE SMALL (9 CASES)	5	3	1	7	1	1
SENSITIVE LONG (4 CASES)	1	3	0	2	2	0
GROSSLY DYSPLASTIC (15 CASES)	10	1	4	10	1	4

TABLE V

Numerical Incidence of Different Nail-Types in Relation to Hand-Types

HAND-TYPE	NAILS				NAILS		
	RUDIMENTARY	PRIMARY	MEDIUM SIZE	LONG	CONCAVE	FLAT	NORMALLY CURVED
ELEMENTARY REGRESSIVE LONG FINGERS (44 CASES)	19	15	5	5	16	10	18
ELEMENTARY REGRESSIVE SHORT FINGERS (26 CASES)	5	13	6	2	3	6	17
ELEMENTARY SIMPLE (5 CASES)	0	1	4	0	0	0	5
LEPTOSOME (12 CASES)	0	2	6	4	3	0	9
SENSITIVE SMALL (9 CASES)	0	1	8	0	2	0	7
SENSITIVE LONG (4 CASES)	0	1	3	0	0	0	4
GROSSLY DYSPLASTIC (15 CASES)	0	7	6	2	0	3	12

TABLE VI

Numerical Incidence of Abnormalities of Crease-Lines and Papillary Ridges in Relation to Hand-Types

ABNORMALITIES OF CREASES AND PAPILLARY RIDGES

HAND-TYPE	SIMIAN LINE	LONG LONGITUDINAL LINE MISSING	CONFUSED CREASE-LINE PATTERN	CREASE LACKING ON DISTAL JOINT, FIFTH FINGER	ARC-TRIANGLE PATTERN FINGER-TIPS	WHORLS OR LOOPS ON PALM
ELEMENTARY REGRESSIVE LONG FINGERS (44 CASES)	5	16	0	0	13	10
ELEMENTARY REGRESSIVE SHORT FINGERS (26 CASES)	4	11	3	1	17	7
ELEMENTARY SIMPLE (5 CASES)	0	0	0	0	0	1
LEPTOSOME (12 CASES)	0	1	0	0	0	0
SENSITIVE SMALL (9 CASES)	1	3	1	0	3	3
SENSITIVE LONG (4 CASES)	0	0	0	0	2	2
GROSSLY DYSPLASTIC (15 CASES)	0	0	0	0	2	2

TABLE VII

Numerical Incidence of Characteristics of the Finger-tips in Relation to Intelligence Quotient

I.Q.	OVER-LONG FINGER-TIPS	OVER-DEVELOPED BUMPS OF SENSITIVITY	BADLY DIFFERENTIATED JOINT	SPATULATE FINGER-TIPS	RUDIMENTARY OR PRIMARY NAILS	WELL-DIFFERENTIATED JOINT	FLAT BUMPS OF SENSITIVITY	NORMAL NAILS
Under 60 (16 cases)	12	12	15	11	12	1	4	4
Between 61 and 65 (40 cases)	16	16	25	15	21	15	10	19
Between 66 and 70 (22 cases)	7	9	12	8	9	10	8	13
Between 71 and 75 (17 cases)	4	6	3	4	7	14	10	10
Between 76 and 80 (11 cases)	2	2	0	2	0	11	9	11
Between 81 and 90 (7 cases)	0	1	0	1	2	7	6	5
Between 90 and 100 (2 cases)	0	0	0	0	0	2	2	2

The rarer hand-types were the elementary simple and the sensitive long. The former belonged to boys who were dull and stable, the latter to psychasthenics with marked anxiety symptoms.

The shape of the finger-tips is part of the whole structural design of the hand and shows for this reason a positive correlation with the hand-type. This is most definitely so in the elementary regressive type of hand either with long or small fingers, where the spatulate finger-tip occurred in a great majority of cases. The same holds good for the leptosome type of hand. Tapered finger-tips were characteristic of the group of grossly dysplastic hands, while a predominance of square terminal phalanxes was found in the elementary simple and sensitive small hands.

DISCUSSION OF TABLES III–VII

TABLE III

A. GENERAL CHARACTERISTICS

It is not surprising that the hypopituitary infantile hand was most often found in the group of boys investigated, as psycho-physical infantilism is very frequent in the unstable, be he mentally defective or not. The hand diagnosis of hypopituitarism could easily be verified by the fact of sexual under-development which involved in most cases both primary and secondary sex characteristics.

Cases of hypothyroic hands were rare. This can be explained by the fact that most of the boys were high-grade defectives.

I have classed in a group by itself those hands which were not straightforward 'mongolian', but had one or two of the typical marks of mongolism. In all these cases the hand alone revealed an endocrine disorder which was neither apparent in other physical trait nor in clinical symptoms. The 15 boys with mongoloid hands had marked extravert tendencies and were highly suggestible. The three hypopituitary-hypothyroic hands belonged to stout but undersized boys with thick-set legs, who in their temperament were stable, obedient and placid.

The hand-diagnosis of eunochoidism could be easily verified by the complete infantilism of the sexual organs apart from the general appearance, which in two cases was, however, atypical.

All types of endocrine hands enumerated in this table were associated with congenital endocrine dysfunction with the exception of the three cases of hyperthyroidism. There was no clinical proof at hand to verify the diagnosis, as none of these boys had had their basic metabolic rate tested, but all of them showed slight exophthalmus. Their hands were the very opposite of hypothyroic hands. They were long and bony hands, the basal phalanxes of which were so thin as to give them a concave shape.

B. PARTICULAR CHARACTERISTICS

1. The abnormal shortness of the fingers was not confined to mongolism but occurred in a high percentage of other forms of endocrine underfunctioning with the exception of hypogonadism.

2. The same holds good for the particular shortness of the thumb and the fifth finger.

3. Hyperflexibility of the hand has also been found to be much more widespread than has hitherto been recognized. It is interesting to note that the flexibility of the palm and of the fingers differed widely. This difference was most conspicuous in the cases of hypopituitary obesity where the finger-tips were hyperflexible and the palms were stiff with the exception of one case.

A discussion of Table IV is unnecessary.

TABLE V

The most interesting result derived from this table is the conspicuous predominance of the under-developed nail in the hands of high-grade boys. The frequent occurrence of this type of nail in mental defectives has been mentioned in earlier writings (1942, 1944). The table shows further that the under-developed nail occurs most often in the elementary regressive hand.

The elementary simple, the leptosome and the sensitive types of hand had the greatest majority of normal nails. It is interesting to note that these hands belonged to boys with relatively high I.Q.'s. A separate account has been made of concave and flat nails, as both had been observed in earlier studies as characteristic of mental defectives. The elementary regressive hand with long fingers showed the greatest number of concave nails, while flat nails were equally frequent (comparatively) in the elementary hand with short fingers.

Pardo-Castello attributed the concave nail, generally called spoon nail, to a neurasthenic disposition, and the same author mentions the rudimentary nail as a symptom of family degeneracy.

TABLE VI

Among the gross abnormalities of the crease-lines the frequent lack of the long longitudinal line is striking. The elementary hand with short fingers is the most abnormal in this respect, followed by the same type of hand with long fingers. This crease-line is sometimes missing in normally intelligent people but only in about 5 per cent.

The simian line was, with the exception of one case, confined to the elementary regressive hand. The simian line is a well-known mark in the hands of mongolian imbeciles, but it occurs also in normally intelligent people in about 2 per cent. I have observed that this line runs in families, which points to its genetic significance.

A confused crease-line pattern has been found to be characteristic of the hands of mongolian imbeciles (see first part of this chapter). It was present in four cases of the boys investigated: three times in elementary regressive hands with short fingers and once in a sensitive small hand. The I.Q.'s of these boys were 60 in two cases, and 58 and 47 in the others.

Papillary Ridges

The papillary ridges have been studied more extensively than the crease-lines, especially in regard to mental deficiency. Cummins (1939), among others, noted that the frequency (1) of arc triangle patterns on the finger-tips, and (2) of loops and whorls on the palmar surface, was much higher in mongoloid than in normal hands. I found the same in mental defectives without mongolism. In this investigation abnormal patterns of the papillary ridges, both on finger-tips and palm, were most frequent in elementary regressive hands with short fingers.

TABLE VII

No investigator who has studied the hands of mental defectives will have failed to notice that their fingers have a peculiar physiognomy. They differ from normal hands not only in their clumsy movements but also in their awkward shape. This table shows that there is a correlation between characteristics of the finger-tips and mental deficiency. Items 1–5 (inclusive) are the most typical of mental deficiency and are generally found together in the same hand. It must be pointed out that the presence of one symptom of mental deficiency in a hand would be misleading for diagnosis. The sum total of hand-traits has always to be taken into account. One finds, for example, spatulate finger-tips in the hands of highly intelligent people, and they are also frequent in the hands of schizophrenics. A short explanation is necessary of what is meant by 'badly differentiated joints'. In a number of hands the creases on the distal joints of the fingers were only faintly marked which caused the impression that the terminal phalanxes were not properly differentiated from the middle phalanxes.

I have attempted in the description of my studies on the mental defective in general, and on the high-grade with delinquent tendencies in particular, to give as precise and definite an information as possible, and for this reason I have quoted large parts of my two papers (1944 and 1947) which deal with this subject. The lay reader may find the large number of tables and the frequent use of technical expressions tedious, but they could not be dispensed with, as only clinical and statistical studies assure any real value for the study of the diagnostic possibilities of hand-features. This cannot be underlined strongly enough. There are a number of people who try to establish a pseudo-scientific interpretation of the hand with an air of medical and psychological 'knowledge', borrowed from medical and psychological books. These people, who are neither physicians nor psychologists, naturally try to cover their scientific shortcomings through quotations from others and through imitation of those with a scientific background of the study concerned. It is almost unnecessary to point out again, that any attempt at hand-diagnosis as I have tried to establish it should only be made by medically and psychologically trained persons.

In the last paragraph of this chapter I am quoting verbally the results and conclusions arrived at from the study of the Colony cases, as I could not express them better and more concisely than I have done in the paper concerned.

1. The group of boys investigated showed a predominance of abnormal hands in regard to both (*a*) general structure, and (*b*) single hand-traits.

2. The majority of the hands were of the elementary regressive type either with long or with short fingers, and these hands showed a great number of similar hand-traits. One can conclude from this that in an abnormal but homogeneous group of individuals there is a decrease of individual variation of hand-traits.

3. The high percentage of endocrine hands in general and of

hypopituitary infantile hands in particular supports the assumption that (*a*) mental and emotional deficiency is linked with endocrine deficiency and (*b*) physical and emotional infantilism appears to be most frequently correlated with an under-developed pituitary gland.

4. Certain single-hand features, such as the degree of flexibility of the palm and fingers, abnormal shortness of the fingers in general and of the fifth finger in particular, are correlated with a number of endocrine disorders.

5. The shape of the fingers, the form of the nails and the degree of flexibility are up to a point correlated with hand-types.

6. Gross abnormalities of the creases and papillary ridges are conspicuously often found in the hands of the boys investigated and occur most frequently in the elementary regressive types of hand and in grossly dysplastic hands.

7. Certain characteristics of the fingers and especially of the finger-tips have been found typical of mental deficiency. These results make it possible to establish a list of marks of constitutional deficiencies as seen in the hand.

MARKS OF EMOTIONAL DEFICIENCY, AS SEEN IN THE HAND

All types of congenital endocrine dysfunction appear to influence emotional reactions, especially hypopituitarism and eunochoidism. The corresponding hand-types are therefore indicative of emotional deficiency. This statement is borne out in detail by everything I said about the correlations between the endocrine hand-types and temperament and mentality in the second chapter of this book.

THE LACK OF EMOTIONAL BALANCE IS INDICATED BY:

1. GENERAL HAND-TYPE:

(*a*) *Elementary regressive hand with short fingers.*
(*b*) *Grossly dysplastic hands.*

2. SINGLE HAND-FEATURES:

(*a*) *Presence of the simian line.*
(*b*) *Long longitudinal line missing.*
(*c*) *Rudimentary or primary nails.*

THE POVERTY AND INHIBITION OF EMOTIONAL RESPONSE IS INDICATED BY:

1. GENERAL HAND-TYPE:

(*a*) *Elementary regressive hand with long fingers.*
(*b*) *Long sensitive hand.*
(*c*) *Marked asymmetry of right hand and left.*

2. SINGLE HAND-FEATURES:

(*a*) *Distal part of the palm flat and narrow.*
(*b*) *Long and thin fingers with overlong terminal phalanxes.*

PERVERSION OF EMOTIONAL RESPONSE IS INDICATED:

GENERAL HAND-TYPE:

(*a*) *Acromicric hand (infantilism with paranoiac trends).*
(*b*) *Eunochoid hands (homosexuality with paranoiac trends).*

CHAPTER VII

THE HAND IN MENTAL ILLNESS

WHILE I was writing this chapter a paper 'Ueber konstitutions-biologische Beobachtungen and den Haenden psyschisch abnormer Persoenlichkeiten' (Observations on the hands of psychopathic personalities), by Leonhard Schlegel, was sent to me by the author. It could not have arrived at a more appropriate moment. It is an interesting paper in several respects:

First of all, it shows that Swiss psychiatrists seriously inquire into the diagnostic significance of hand-traits. The paper contains the results of an investigation on the hands of inmates of the 'Kantonalen Heil- und Plegeanstalt Rheinau' (Superintendent Professor Hans Binder).

Secondly, the paper shows that apparently a scientific *frontière* exists between certain Swiss psychiatrists and those of France, England and America, as only few other than Swiss and German authors are mentioned in the bibliography, attached to an investigation that was made at a Mental Hospital, whose Superintendent is well known outside the Swiss borders. There are only 8 out of 84 references which are either English, American or French; 75 references are German. This is a regrettable state of affairs, though I was personally pleased to find some recent German publications mentioned, which had not been accessible to me during the war years. The earlier German works, referred to by Schlegel, have been already mostly mentioned in my book *The Human Hand*. They concern, with one exception, the study of the papillary ridges.

Schlegel did obviously not know my book *The Human Hand* nor my other writings as the only authors he refers to, as having done work on constitutional hand-traits in mental illness are Friedemann and Kuehnel. Schlegel describes the latter as a pupil of Kretschmer and says that his research was entirely orientated on Kretschmer's physical types. Schlegel himself has followed the same line of research. He does not know of Sheldon's work, which in most ways has outmoded that of Kretschmer, though the latter's merit as a pioneer remains unaltered. For this and other reasons Schlegel's investigation is not up to date.

He is hampered in both the general direction of study and the interpretation of detailed findings as well as by his lack of knowledge of previous research work. For this reason the results reported in his paper are of little value.

Apart from this very serious reservation, a few statements he makes are worth discussing:

Schlegel adheres to the idea that certain abnormalities of the form, the creases and the papillary ridges of the hand, can be considered degenerative stigmata and he gives a sound and careful definition of the latter, which I translate here:

> Degenerative stigmata are physical deviations from the norm, which are visible, congenital and constitutional and which occur relatively rarely. They have little functional significance and are not determined by gross pathological changes and are frequently met with in individuals, both physically and mentally abnormal, and also in their relatives.

While Schlegel does not contribute anything new to the interpretation of the papillary ridges of the hand in relation to mental illness, he points out an interesting fact about the origin of the main crease-lines in man, a fact that I had already mentioned in *The Human Hand*, namely that the main transverse creases and the thenar line are not due to hand movements, but are inherited and therefore part of a prenatal constitutional design.

The origin and design of the accessory crease-lines and probably also of the long longitudinal line are largely determined by psychomotor habits of the hand.

Schlegel goes into detailed discussion about the simian line (*Vierfingerfurche* in German) and mentions that this line is frequently found in imbeciles and mental patients; a fact which is already known. He also mentions that faults of the main creases are more frequently found in mental patients than in normal persons. These findings correspond to research done earlier by other authors, namely E. Wilson and myself. He mentions a more recent study by a German, Miss Schiller (*Zeitschrift fuer menschliche Vererbung und Konstitutionslehre*, 1942) which shows that:

1. CERTAIN MAIN CREASE-LINES ARE HEREDITARY CREASES, and
2. THAT A DISHARMONIOUS PATTERN OF CREASE-LINES IS FAR MORE FREQUENT IN MENTAL PATIENTS THAN IN NORMAL INDIVIDUALS.

These findings also correspond to my own observations, reported in my earlier writings as well as in this book.

The interest of Schlegel's paper is not in his findings, but in his inquiry into the whole question of the diagnostic significance of hand-traits as well as in the fact, that is evident from the references given, that certain detailed observations have been made independently in different countries.

Kretschmer had made the study of mental patients the point of departure of his well-known typology (Koerperbau und Charakter). He came to conclusions about the constitution of the normal individual through studying the abnormal. This procedure is open to criticism, if the results are not checked carefully and compared in a great number of cases with data about the psycho-physical make-up of ordinary people. There is, however, much to be said for Kretschmer's approach. Mental patients have practically no control over their emotional reactions and expressions and offer therefore a direct insight into the working of their minds.

To study only the hands of mental patients in order to get an idea of the correlations between hand-traits and personality in general would have been an insufficient investigation and would have been out of touch with reality. Not to study these hands at all would have been a grave omission and would have left a gap in the knowledge of similarities as well as deviations between the hands of normal and psychotic people. I have therefore studied the hands of the abnormal personality side by side with those of the normal, but have given preference to the systematic study of mental patients for several reasons:

1. The study of the hands of mental patients was directly linked with the preceding investigation on the hands of mental defectives.

2. Extreme qualities bring essential features always more to the surface and into relief than moderate ones—and in mental patients we are dealing with extreme qualities of the mind.

3. Every form of mental illness or of any psychological disturbance goes with a set of symptoms which are common to every patient, and this similarity facilitates research. A valid study needs therefore a much smaller number of cases than any investigation on normal individuals.

4. Psychology and psychiatry are so closely linked that they present practically the same branch of medicine: the inquiry into the psychological diagnosis includes automatically inquiry into the psychiatric diagnosis as well.

I have been able to study the hands of mental patients extensively and to collect material of patients suffering from

1. SCHIZOPHRENIA
2. MANIC-DEPRESSIVE PSYCHOSIS
3. PARAPHRENIA
4. EPILEPSY

A. THE HANDS OF SCHIZOPHRENIC PATIENTS

A short note on schizophrenia will be useful to the general reader, as the notion which the layman has about schizophrenia or split personality is rather confused.

Schizophrenia means division of mind. The term split personality is therefore well chosen, but must not be confused with that of 'double personality,' which designates a person with contradictory attitudes of behaviour.

The principal characteristics of schizophrenia are the loss of the feeling of identity together with the loss of emotional vitality. The split is in two regions of the personality: the mental and the emotional one.

The loss of identity leads to identification with ideal images and in consequence to an illusory assumption of 'power'. Personalities such as Christ, the Emperor of China, the Virgin Mary, the Queen of Spain are generally chosen and the whole behaviour of the patients concerned is modelled according to their ideas of the 'Great Person' they represent.

The loss of emotional vitality leads, naturally, to contactlessness and to the return to the autistic-narcissitic phase of childhood.

Present-day psychiatry does no longer recognize a clear-cut division between the three principal psychoses: schizophrenia, manic-depressive psychosis and paraphrenia, though an essential difference exists between the first and the second. Manic-depressive psychosis is mainly an illness of the emotions. The emotional reactions are accentuated in both directions, that of pleasure and of displeasure. Especially in the hypomanic phases the pleasurable emotions govern the mind and put the person concerned into a state of euphory, while the opposite occurs in the phases of depression. There are some cases with a mixed psychosis, showing both syndromes, that of schizophrenia and of manic-depressive psychosis. One always finds in paraphrenia elements of schizophrenia and often elements of manic-depressive psychosis. The main characteristic of paraphrenia is the obsessional content of ideas. The obsessed personality is both anxious and aggressive and both features are always found in paraphrenic patients.

Schizophrenic psychosis can take different forms and since Kraepelin four types of schizophrenia have been distinguished:

1. Schizophrenia Simplex

which corresponds on the whole to the older notion of dementia precox; the accentuation is here on the progressive mental deterioration.

2. Hebephrenia

represents a syndrome which goes with mannerisms, as all forms of schizophrenia do, and with hallucinations of a predominantly pleasant type. Hebephrenics are generally found among the younger patients. Deterioration of human faculties concerns both the sphere of the instincts and of the intelligence. Hebephrenics regress in their habits to the 'animal' level. They have very bad eating manners and are completely uninhibited in their greed for food. Their physical build and their hands are, as a rule, coarse.

3. Catatonia

In catatonic patients motor symptoms are in the foreground. They are grimacing and grinning all the time as if their minds were under the influence of a constant cramp. They have an extraordinary perseveration of postures, and they give, just like people under hypnosis, a vivid illustration of the strange 'powers' of the mind over all known physiological laws. Catatonic patients hold their arms, for example, upright or bent backwards for hours on end without any sign of fatigue.

4. Paranoid Schizophrenia

As the term indicates, ideas of reference and persecution colour the mental imagery. It is often difficult to distinguish the paranoid schizophrenic from the paraphrenic patient. The differential diagnosis relies on the more severe disintegration of the schizophrenic, who has practically never a 'normal' moment, while the paraphrenic with ideas of persecution behaves at times quite normally. He is the person with a kink, while the schizophrenic is altogether removed from any experience of reality.

I have shown in *A Psychology of Gesture*, that the strange mannerisms of the schizophrenic patient can be explained as being obsessions which are kinaesthetically expressed. The subconscious disturbances in schizophrenia are less manifest in the emotional and mental sphere than in the psycho-motor sphere. In no other form of mental illness does one find the same quantity and quality of mannerisms.

I have seen schizophrenic patients at St. Bernard's Hospital and also at St. Lawrence's Hospital at Caterham.

According to the material that I have collected, schizophrenia is by far the most frequent psychosis and this corresponds to the general

TABLE I

deals with the Numerical Incidence of Certain Hand-Types in Relation to the Different Forms of Schizophrenia, in 245 Patients (175 Women and 70 Men)
(m. = male; f. = female)

HAND-TYPE	SEX	PARA-NOID TYPE OF SCHIZO-PHRENIA	CATATONIC TYPE OF SCHIZO-PHRENIA	HEBE-PHRENIC TYPE OF SCHIZO-PHRENIA	SCHIZO-PHRENIA SIMPLEX	SECONDARY DEMENTIA	TOTAL
ELEMENTARY REGRESSIVE	f.	2	3	5	10	20	80
	m.	3	15	7	15		
ELEMENTARY SIMPLE	f.	0	0	1	2	0	3
	m.	0	0	0	0		
BONY-MOTOR	f.	5	4	3	3	2	19
	m.	2	0	0	0		
FLESHY-MOTOR	f.	0	3	4	2	3	15
	m.	0	2	0	1		
SENSITIVE LONG	f.	21	16	0	3	5	60
	m.	8	3	2	2		
SENSITIVE SMALL	f.	40	0	3	3	4	58
	m.	4	2	0	2		
ACROMICRY	f.	4	0	0	1	3	10
	m.	2	0	0	0		
TOTAL		91	48	25	44	37	245

TABLE II

deals with the Numerical Incidence of Certain Characteristics of the Finger-tips and Nails, in relation to Size and Mobility, as found in the Different Hand-Types, in 245 Schizophrenics (175 Women and 70 Men)
(m. = male; f. = female)

HAND-TYPE	SEX	HUMAN DIGITAL FORMULA	SIMIAN DIGITAL FORMULA	FIFTH FINGER TOO LONG	FIFTH FINGER TOO SHORT	EQUAL LENGTH THIRD AND FOURTH FINGERS	THUMB TOO SHORT	THUMB TOO LONG	ANKYLOSIS OF FIFTH FINGER	ANKYLOSIS OF ALL FINGERS
ELEMENTARY REGRESSIVE WITH LONG FINGERS (53 CASES)	f.	5	25	17	0	6	6	7	3	0
	m.	5	18	15	0	12	2	0	1	0
ELEMENTARY REGRESSIVE WITH SHORT FINGERS (27 CASES)	f.	0	10	0	7	0	0	0	0	0
	m.	3	14	0	6	0	0	0	0	0
ELEMENTARY SIMPLE (3 CASES)	f.	0	3	2	0	0	0	0	0	0
	m.	0	0	0	0	0	0	0	0	0
BONY-MOTOR (19 CASES)	f.	13	4	8	2	2	1	0	0	0
	m.	0	2	1	0	1	0	0	0	0
FLESHY-MOTOR (15 CASES)	f.	2	10	3	1	3	1	0	0	1
	m.	2	1	1	1	1	0	0	0	0
SENSITIVE LONG (60 CASES)	f.	23	22	24	3	11	1	0	0	0
	m.	1	14	10	1	8	0	1	0	0
SENSITIVE SMALL (58 CASES)	f.	25	25	23	8	11	1	2	1	0
	m.	1	7	5	2	2	0	0	0	0
ACROMICRY (10 CASES)	f.	1	7	0	2	1	0	0	0	0
	m.	0	2	0	0	0	0	1	0	0

TABLE III

deals with the Numerical Incidence of Certain Characteristics of the Finger-tips and Nails in the Hands of 245 Schizophrenics in Relation to the Different Forms of Schizophrenia

(m. = male; f. = female)

FORMS OF SCHIZOPHRENIA	SEX OF PAT.	FINGER-TIPS					NAILS				MISSHAPEN	
		SPATULATE	CONICAL	SQUARE	OVER-LONG	HYPER-FLEXIBLE	LONG	MEDIUM	SMALL	UNDER-DEVELOPED	FIFTH FINGER	THUMB
PARANOID (91 CASES)	f. m	20 3	27 10	25 6	16 3	35 13	19 4	30 7	15 6	8 2	5 0	2 1
CATATONIC (48 CASES)	f. m.	18 13	3 7	5 2	5 9	8 5	2 8	11 11	9 3	4 0	6 1	3 2
HEBEPHRENIC (25 CASES)	f. m.	8 5	6 0	2 4	6 6	7 1	1 1	6 5	3 2	6 1	2 0	2 1
SCHIZOPHRENIA SIMPLEX (44 CASES)	f. m.	14 5	3 5	7 10	2 0	9 4	3 2	4 7	4 5	13 6	6 4	8 6
SECONDARY DEMENTIA (37 CASES)	f.	18	5	14	4	10	2	14	15	6	17	12

TABLE IV

deals with the Numerical Incidence of Abnormalities of the Crease-Lines and Papillary Ridges in Relation to the Different Forms of Schizophrenia
(m. = male; f. = female)

TYPE OF SCHIZOPHRENIA	SEX	*ABNORMALITIES OF CREASE-LINES*					*ABNORMALITIES OF PAPILLARY RIDGES*			
		SIMIAN LINE	LONG LONGI-TUDINAL MISSING	CONFUSED CREASE-LINE PATTERN	FAULTS IN UPPER TRANSVERSE LINE	FAULTS IN LOWER TRANSVERSE LINE	WHORLS OR LOOPS ON THENAR EMINENCE	WHORLS OR LOOPS ON HYPOTHENAR EMINENCE	ARC-TRIANGLE PATTERN ON ONE OR TWO FINGER-TIPS	ARC-TRIANGLE PATTERN ON ALL FINGER-TIPS
PARANOID	f.	4	0	2	6	15	0	9	8	2
	m.	1	0	0	3	1	1	0	4	0
CATATONIC	f.	4	0	0	4	10	2	10	7	1
	m.	2	1	0	2	2	0	0	1	0
HEBEPHRENIC	f.	3	0	0	5	4	0	2	3	1
	m.	1	0	0	0	0	0	0	0	0
SCHIZOPHRENIA SIMPLEX	f.	6	1	11	9	15	1	5	12	5
	m.	9	2	1	9	9	1	3	13	0
SECONDARY DEMENTIA	f.	7	0	4	7	10	0	7	11	2

trend of our civilization towards disintegration. The disintegrating factor, so blatant in our present-day Society, affects on one hand each individual, and represents on the other hand the sum total of the more or less complete break in Man's soul. This statement does, of course, not imply that other psychoses are not illnesses of disintegration, but in schizophrenia the split is most complete. Schizophrenia may be called the typical psychosis of our time for this reason:

Our Age is that of intellectual advance but also of emotional impoverishment. We are intellectually more grown up than other generations, but as wisdom and emotional vitality do not go hand in hand with this development, disintegration of the 'collective' as well as the 'individual' mind is the natural consequence. This might explain the higher frequency of schizophrenia than of manic-depressive psychosis which goes with a more localized disintegration, affecting in the first place the emotions.

I regret that my material consists of a far larger number of the hands of schizophrenics than of other psychotic patients, as an equal number of every category would have been an ideal condition.

I have investigated the hands of the schizophrenic patients according to the four different types of schizophrenia and have set out to show certain correlations which exist between those and hand-traits by means of four tables.[1]

DISCUSSION OF THE TABLES

The material of study relates to 245 schizophrenic patients, 175 women and 70 men. Their ages ranged from 18 to 65 years, the majority of both sexes being between 35 and 45 years of age.

TABLE I

Table I shows the numerical incidence of different hand-types in relation to the different forms of schizophrenia.

Two hand-types are definitely characteristic of schizophrenia:

1. THE ELEMENTARY REGRESSIVE HAND
2. THE SENSITIVE HAND OF BOTH VARIETIES (SENSITIVE LONG AND SENSITIVE SMALL).

By far the most frequently found hand is the elementary regres-

[1] I have added to the four principal groups of schizophrenia 37 cases of secondary dementia. This is a form of schizophrenia which resembles imbecility or idiocy and which is met with in patients of an advanced age.

sive which occurred in 80 cases. It seems particularly characteristic of (*a*) schizophrenia simplex, (*b*) catatonia and (*c*) hebephrenia. The fact that 20 out of 37 women with secondary dementia possessed this type of hand seems to me less characteristic for schizophrenia than for a constitutionally 'low' mentality; most of these women had been feeble-minded before the onset of schizophrenia.

The sensitive hand, and particularly the sensitive small one, is undoubtedly characteristic of the paranoid type of schizophrenia. 44 patients out of 91, who suffered from this form of the illness, possessed this type of hand: 40 were women and 4 men. The great difference in number reflects not only the fact that there were far less male than female patients under investigation but also shows that the sensitive small hand is really a woman's hand and occurs rarely in men.

The sensitive long hand was found in 60 of all cases and in 29 of all those with the paranoidal type of the illness. 21 of these were women and 8 were men.

It is also interesting to note that acromicry seems to be correlated with the paranoid type of schizophrenia, as the majority of cases, namely 6 out of 10, were found in this condition. The acromicric hand can be considered a miniature edition of the sensitive small one and designates an endocrine dysfunction: namely hypopituitary infantilism. One may assume that paranoidal schizophrenics have altogether a tendency towards endocrine disturbance of a hypopituitary type, as the sensitive small hand may be considered (see Chapter II of this book) a symptom of hypopituitary infantilism. The sensitive long hand, especially in its very marked form, denotes a tendency to hypogonadism, an endocrine dysfunction, which has been associated in modern psychiatric research with schizophrenia. It is well known that many schizophrenic men are 'feminine' in appearance and homosexuality is often found among them.

The bony motor hand takes the fourth place in frequency of hand-types in schizophrenia. It occurred in 7 cases in paranoid schizophrenics. The bony motor type of hand is similar to the long sensitive one. It is, so to speak, a more virile edition of the latter. One finds the same elongated shape of palm and fingers, the difference being only the stronger bony structure and better muscular development; the latter showing a more active and vital disposition.

Both the sensitive long and the bony motor type of hand also occurred to a significant degree in catatonic schizophrenics.

The fleshy motor hand, which one may call the most 'normal'

and the most fully developed type of hand, was found most often in the hebephrenic type of schizophrenia. This can be easily understood by what I have said about the mental and instinctual concomitants of this form of the illness. Hebephrenic patients are more 'animal' than the other schizophrenics, but appear much more normal in their behaviour as they exhibit much less mental torture and often even a certain *joie de vivre* which reminds one of hypomanic patients. As we shall see later, patients with manic-depressive psychosis have elementary and fleshy motor hands in a significant number.

Elementary simple hands are a simplified edition of the fleshy motor hand and are very rarely found in schizophrenics. It is evident from the results of this table that certain hand-types are significantly often found in the different forms of schizophrenia. The fact that the elementary regressive type of hand appears most often in schizophrenia simplex and in the catatonic form of schizophrenia can be explained by the fact that the former condition very often develops into the latter at a later stage of the illness. A similar situation is found in the hebephrenic patient who frequently becomes catatonic. The paranoidal schizophrenic remains more on the obsessional and altogether 'mental' level, and as the table shows, his hands are of a definitely different shape. This has been, in a more general way, pointed out by Friedemann in his 'Handbau und Psychosis', a paper which is of considerable interest and value. Friedemann investigated (1928) the hands of 55 schizophrenic patients, 30 men and 25 women, as well as those of 45 patients with manic-depressive psychosis, 25 men and 20 women. His inquiry concerned both the shape of the hands as well as many other hand-traits and typical gestures.

He makes the same distinction between the four forms of the illness as I have done and the distribution of his cases in the different groups is the following:

HEBEPHRENIC	10 male	8 female
CATATONIC	10 ,,	8 ,,
PARANOID	5 ,,	5 ,,
SIMPLEX	5 ,,	4 ,,

As Friedemann has not a clear-cut classification of hands it is not possible to compare his findings with mine in a direct way. He speaks of athletic hands in hebephrenic women, which corresponds to my own findings, as the regressive elementary hand is a strong and 'masculine' hand. He actually mentions the word 'masculine' in this connexion. He found on the whole the leptosome and the 'aristocratic' hand, which corresponds to my bony motor and sensitive long hands to be the most characteristic hand-types and after those the 'athletic' one by which he roughly means what I have called the elementary regressive hand. The 'pyknic hand', which corresponds to my fleshy motor and elementary simple types, is the rarest Friedemann found in schizophrenic patients, which again is confirmed by my own findings.

The number of patients investigated by Friedemann is very small and his designation of hand-forms is vague. Apart from this reservation his study has value, particularly as it is the most detailed of its kind, which has come to my knowledge and also from the fact that he has done pioneer work in regard to the diagnostic significance of the hand in relation to mental illness.

DISCUSSION OF TABLE II

This table deals with the numerical incidence of certain characteristics of the fingers in relation to different hand-types in 245 schizophrenic patients. Most characteristics investigated concern the size of the fingers.

THE DIGITAL FORMULA

I have in the first place investigated the frequency of the human and the simian digital formula in relation to hand-types. In people with a human digital formula the index finger is longer than the fourth digit and with those with a simian digital formula the position is reversed. It is common knowledge that the latter is by far more frequent among 'normal' people. This is also the case in schizophrenic patients, but as a whole in them the human digital formula occurs more often than in other human beings.

The human digital formula was most often present in relation with the bony motor hand; out of 19 patients with such a hand, it was found in 13 cases. As the 'bony' hand is characteristic of the paranoid type of schizophrenia, one may conclude that the human digital formula has a special relevance for this type of the illness. This conclusion is supported by the fact that it occurred in 26 out of 58 sensitive small hands and in 24 out of 50 sensitive long ones, both of which being equally characteristic of paranoid schizophrenia. I have distinguished in this table two varieties of the elementary regressive hand, one with long and one with short fingers. The human digital formula occurred significantly more often in the first variety, which is by far the most frequent in schizophrenics. The human digital formula also occurred in fleshy-motor hands, but was never present in the elementary simple hands investigated.

THE FIFTH FINGER

Special attention has been paid to the length of this finger for the following reason: The marginal fingers, which are the thumb and the little finger, are much more varied in size and shape than the other digits and it has been observed that in endocrine dysfunction both, but especially the latter, show special characteristics of shape and size. The abnormally short fifth finger in mongolian imbeciles is one of the well-known marks of the condition. During my investigation of schizophrenics, I had been observing that the fifth finger was often unusually long, and this observation is fully confirmed by the statistical results obtained.

An overlong fifth finger was most often found in the elementary regressive hand with long fingers with 32 out of 53 cases, then in the sensitive long hand where it occurred in 34 out of 60 cases, then in the sensitive small hand where it was found in 28 out of 58 cases, and after that in the bony motor type of hand with 9 out of 19 cases. The overlong fifth finger seems to be particularly characteristic of schizophrenia of all varieties, though it seems to have special predominance according to my documents for the catatonic and paranoidal types of the illness.

A too short fifth finger is of much rarer occurrence and is mostly related to the sensitive small and the elementary hand with short fingers. In both hand-types, which are, as their very name indicates, short hands, this symptom has probably an endocrine significance.

EQUAL OR ALMOST EQUAL LENGTH OF THE MEDIUS AND FOURTH FINGERS

While studying my material, I was struck by the unusual digital formula which this abnormality represents, as I have never seen these two fingers being equally long in any normal person. I want to mention here that the physiognomy of the hands of schizophrenics strikes one at first sight by the elegance and beauty of their fingers, which are in the majority of cases very long, slim and straight and it is no wonder that Friedemann speaks in this connexion of aristocratic hands. The equal length of the third and fourth fingers occurs mostly in long hands. This symptom was present in 18 out of 53 elementary regressive hands with long fingers, in 19 out of 60 in the sensitive long and in 13 out of 58 sensitive small hands. It occurred 4 times in fleshy- and 3 times in bony-motor hands.

As this abnormality is very rare in the hands of other psychotic or mentally defective patients, it must therefore be regarded as a specific mark of schizophrenia.

THE THUMB

Deviations from the normal in regard to the size of the thumb appears to be associated in the first place with endocrine dysfunction, and secondarily with mental deficiency.

One may expect that the thumb will be long in hands with long fingers.

The position is, however, not as simple as this table shows: the thumb was too long in 7, and too short in 8 cases in the elementary type with long fingers. On the whole, a deviation from the normal in thumb-length is an unusual symptom in the hands of schizophrenics and therefore of no diagnostic importance. The fact that I found practically an equal frequency of too long and too short thumbs in the elementary hand with long fingers only proves the dysplastic character of this type of hand.

ANKYLOSIS

Ankylosis of all fingers is a rare deformity and has been recognized as a degenerative stigma, sometimes found in mental defectives. Ankylosis of the fifth finger is also well known as a degenerative stigma in psychotic patients. The latter occurred in 5 cases, the former in one case. It is noteworthy that 4 patients with ankylosis of the fifth finger had elementary regressive hands with long fingers, another proof of what I have already pointed out at several occasions, namely that most symptoms of genetic significance are found in this type of hand.

DISCUSSION OF TABLE III

This table deals with the numerical incidence of certain principal characteristics of the finger-tips and nails in relation to the different forms of schizophrenia. I have shown in *The Human Hand* that the different hand-types are correlated with different shapes of the fingers. Finger-tips are either spatulate, conical, square or pointed. The latter form is rare and I have not taken it into account in this investigation. I must mention here that the five fingers are practically never of exactly the same shape; the index and fifth finger have always the tendency towards a conical form and, in fact, most pairs of hands have mixed finger-tips. One form or the other is, however, in most cases predominant, and the distinction of either spatulate, conical or square finger-tips has been made accordingly.

Among normal people the spatulate finger-tip is, as far as my observations go, the rarest and the square one the most frequent.

The spatulate finger-tip occupies the first place in all forms of schizophrenia except the paranoid and may be considered as an important attribute of the 'schizophrenic' hand. It is mostly associated with the elementary regressive hand-type. According to this table, it occurred most frequently in catatonic schizophrenia, then in schizophrenia simplex and hebephrenia, after that in a comparatively much smaller number in paranoid schizophrenia. This distribution can easily be understood from the results of the previous tables, which show the predominance of elementary regressive hands either with long or short fingers in all forms of schizophrenia except the paranoid one.

The conical finger-tip is predominant, as one would expect, in paranoid schizophrenics, where it occurred most frequently, and is comparatively rarely found in other forms of schizophrenia.

The square finger-tip is more frequently found than the conical one in schizophrenia simplex and secondary dementia.

THE OVERLONG FINGER-TIP

I have defined in the preceding chapter what I understand by an overlong finger-tip and need not repeat the definition here. I have also already pointed out that the fingers of schizophrenic patients are mostly of unusual length, very straight and beautifully shaped. One can expect that long hands with long fingers will have very long finger-tips, but what I mean by 'overlong' finger-tips does not seem to be necessarily correlated with long hands and long fingers only. I also found it in a great many cases in the hands of mental defectives, and most often in the elementary regressive hand. This hand-trait seems to be characteristic of both mental deficiency and a schizoid make-up.

THE NAILS

I have correlated the size of the nail with the different types of schizophrenia. Of course, one finds long or very long nails mostly in long sensitive and bony motor hands and short ones in broad hands.

As with normal people the medium-sized nail takes first place also in all types of schizophrenia. The very long nail is mostly found in paranoid schizophrenics, who have frequently long, sensitive and bony motor hands furnished with this type of nail. The small nail is most often found in secondary dementia, after that in catatonic schizophrenia, and the under-developed nail occurred most often in cases of schizophrenia simplex and hebephrenia. It was rare in paranoid and catatonic patients. The under-developed nail is characteristic of the elementary regressive hand, especially of the short variety, which is so predominant in mental defectives, where I have observed the undeveloped or rudimentary nail in a great many cases, as pointed out in the preceding chapter.

THE FLEXIBILITY OF THE FINGER-TIPS

Among the physical qualities of the hand the degree of flexibility of both the palm and the fingers deserves attention, in regard to diagnosis. I have only mentioned here the flexibility of the finger-tips and I have not dealt here with the flexibility of the palm. It is well known that muscular rigidity and rigidity of the articulations can be induced by a permanent nervous tension. Obviously, the schizophrenic individual is victim of such nervous rigidity, especially in the catatonic form of the illness. The low flexibility of the palm of schizophrenics is therefore an item which is so obvious that it need hardly be mentioned.

The flexibility of the finger-tips seems to be quite another matter, a physical trait of the hand, which deserves a special investigation. My interest in this hand-feature had been particularly aroused since I had worked at the Endocrine Clinic of Dr. Simpson's where I observed that every case of hypopituitary obesity showed hyperflexible finger-tips and at the same time rigid palms. I demonstrated the hands of these patients to Dr. Simpson, who considered this item a characteristic mark of the condition.

The interest of the cases just mentioned lies in the seeming contradiction between the flexibility of the palm and the fingers, which apparently is the result of pituitary dysfunction. One must remember that the terminal phalanxes of patients with hypopituitary adipositas are very short and mostly of a tapered or pointed shape, which is proof of too early a closure of the epiphyses, preventing the full development of the finger-tips.

In schizophrenics where we find also very often the hyperflexible finger-tips, the situation is quite different, indeed opposite. This item was most frequent in the hands of paranoid schizophrenics where it occurred in 48 out of 91 cases. We find this hand-trait also in catatonic patients where one would expect it least. In the long sensitive type of hand one often finds a hyperflexible finger-tip, which in these cases does probably not refer to pituitary dysfunction, but may be due to subgonadic function.

ABNORMALITY OF THE SHAPE OF THE FIFTH FINGER AND THE THUMB

Ankylosis and other deformities of the fifth finger are well-known attributes in schizophrenia and that is why I have put the little finger before the thumb in this discussion.

The most frequent occurrence of a misshapen fifth digit was noticed in patients with secondary dementia, of whom the elementary regressive hand is very characteristic, and it is in this type of hand that one finds abnormal marginal digits most often.

The hand-trait is comparatively rare in paranoidal schizophrenics. This is what one would expect, as the sensitive hand, either long or small, is characteristic of these patients, hands which have, as a rule, well-shaped fingers.

A Misshaped Thumb

This occurred less often than the previous trait. It was most frequently found in patients with schizophrenia simplex and with secondary dementia; both forms of the illness go with types of hands in which degenerative stigmata occur frequently.

The reader will by now be able to discern a pattern of symptoms belonging to certain hand-types. The elementary regressive hand with either short or long fingers, is, according to my experience, abundantly equipped with specific markings, and this hand-type is characteristic of both, the mental defective and the schizophrenic patient, in whom psychosis is superimposed on mental deficiency. The elementary regressive hand is also frequent in hebephrenic patients, who are, as a rule, of a far more 'primitive' fibre than paranoid schizophrenics. In the latter one finds the most developed hands with far less genetic marks, and it is well known that these patients possess an accentuated sensitivity and often preserve their mental capacities up to a point, though out of tune with reality.

DISCUSSION OF TABLE IV

This table refers to abnormalities in both crease-lines and papillary ridges in the different groups of schizophrenia.

1. THE SIMIAN LINE

As the elementary regressive hand is found comparatively often in schizophrenia simplex, the frequent presence of the simian line is a natural consequence. It occurred in patients with this form of schizophrenia in 33·8 per cent, which is a very high percentage, if one remembers that this crease-line only occurs in about 2 per cent in ordinary people.

The fact that 7 out of 36 patients with secondary dementia possessed this abnormal crease-line is not surprising, because one finds the elementary regressive hand fairly often with these patients. The same holds good for catatonic schizophrenics, where the simian line was present in 6 out of 43 cases. In paranoid schizophrenics this hand-trait is comparatively rare.

2. LONG LONGITUDINAL LINE MISSING

This hand-trait occurs rarely in schizophrenics, a fact that can be explained by the general tendency of their hands to be long and narrow rather than short and broad.

3. FAULTS IN THE MAIN CREASES

E. Wilson and L. Schlegel observed the far greater frequency of faults in the main transverse creases in psychopathic than in normal persons and my own findings confirm theirs. The creases in the hands of patients with schizophrenia simplex were the most 'faulty' ones. Those with secondary dementia came second and the catatonic schizophrenics were third. Hebephrenic patients had the most normal crease-line systems.

In regard to detailed findings it is interesting to note that faults in the lower transverse crease-line are found more frequently than in the upper one in all types of schizophrenia, except hebephrenia.

4. A CONFUSED CREASE-LINE PATTERN

the significance of which I have explained earlier in this book, was found comparatively often in patients with schizophrenia simplex, a result that can be easily understood, as this form of schizophrenia often goes with congenital imbecility.

5. ABNORMALITIES OF THE PAPILLARY RIDGES

showed that unusual patterns on the thenar eminence, such as whorls and loops, were very rarely found in all types of schizophrenia, while the same patterns were most frequent on the hypothenar eminence. The catatonic patients investigated possessed this hand-trait most often, namely in 10 out of 48 cases.

Arc-triangle patterns on one or two finger-tips occurred most frequently in schizophrenia simplex and in secondary dementia. This hand-trait is, as I have mentioned several times in this book, particularly connected with the elementary regressive type of hand, and is therefore found in cases either of mental deficiency or of a psychopathic personality. As this pattern of the papillary ridges also occurs in normal hands, the diagnostic value of it seems rather small.

The same pattern on all finger-tips is very rare in normal people and therefore more interesting from a diagnostic point of view. It occurred five times in patients with schizophrenia simplex, and only twice in paranoid schizophrenics and patients with secondary dementia.

HANDS OF PATIENTS SUFFERING FROM PARAPHRENIA, MANIC-DEPRESSIVE PSYCHOSIS AND EPILEPSY

I have already mentioned that I regret not to have been able to collect the material for my hand-studies of mental patients in equal numbers. The hands of schizophrenics represent by far the greater number of all cases and for this reason alone I would have given them a place apart in this chapter.

The cases of other mental illnesses studied were fewer, but the number of patients was equal in the three groups of

PARAPHRENICS
MANIC-DEPRESSIVE PATIENTS and
EPILEPTICS.

The hands of 80 patients were investigated in each category.

I have mentioned briefly in the beginning of this chapter the most essential features which are characteristic of the three psychoses and shall only add a few sentences here about the principal symptoms of manic-depressive psychosis and paraphrenia. The main symptoms of epilepsy are so widely known also to the lay reader that there is no necessity to point them out here. The reader who is particularly interested in this condition, finds a note on epilepsy in my book *A Psychology of Gesture*.

I only want to mention here that epilepsy is frequently found in conjunction with schizophrenic psychosis or imbecility or both.

Manic-depressive psychosis is an illness that affects both the emotions and thought processes. The name implies already the definition and clearly emphasizes the emotional character of the illness. 'Manic' refers to uncontrollable emotions, a state of mania, where all reason is barred and man acts under a tremendous pressure of impulse. 'Depressive' is an attribute of temperament and also concerns the emotions. Depression, as the word implies, means a slow, negative, sad emotional atmosphere with the outcome of hesitation, indecision, withdrawal from human contacts and minimum of activity.

Both states, mania and depression, alternate in the patient and the French have rightly given the name *maladie circulaire* to this type of psychosis. In between the two extreme states of mind one observes an intermediary or hypomanic phase, which may be considered the more normal medium. While any inhibition is removed in the manic state and the patient acts without control, thinks without judgement and sense of reality, he lives in the interim phase of hypomania in an euphoric condition, his violence tempered down by a pleasant, happy feeling of omnipotence and benevolence towards his surroundings. The world appears to be rosy, everything easy and he himself an excellent specimen of the human race. He makes contact most easily and has no grudge, only the 'positive' emotions towards his fellowmen. I don't hesitate to say that the state of hypomania is an enviable one. Most readers will know, even from their milder experiences of depression, how black, inhibited and hopeless everything looks in such a state of mind. While in a state of mania and also, to a minor degree, in hypomania, the mind is engaged in exaggerating and magnifying thoughts about the Self into ideas of grandeur and shows its 'incontinence' in flight of ideas and complete absence of concentration, and behaves so to speak like a river bursting over its borders, quite the opposite processes take place in depression. Ideas of worthlessness, lack of thinking power, poverty of associations, absence of dynamism, both physical and emotional, govern the mind which behaves like a river dried up.

Paraphrenia means a mind that is beside itself or gone astray. I have already mentioned that paraphrenia always goes with an obsessional frame of mind, which affects both the emotions and thoughts. Paraphrenic people suffer from strange twists or kinks. The contents of their obsession are either religious or sexual or both and take most often the form of ideas of persecution. The latter are, indeed, the most frequent complaint of the illness and range from

TABLE IA

deals with the Numerical Incidence of Certain Hand-Types in:

1. *80 paraphrenics (all women)*
2. *80 manic-depressive patients (35 men and 45 women)*
3. *80 epileptics (30 men and 50 women)*

(m. = male; f. = female)

HAND-TYPE	PARAPHRENICS	MANIC-DEPRESSIVE PATIENTS	EPILEPTICS
ELEMENTARY REGRESSIVE WITH SHORT FINGERS	8	m. 3 f. 3 6	m. 8 f. 6 14
ELEMENTARY REGRESSIVE WITH LONG FINGERS	4	m. 3 f. 3 6	m. 10 f. 6 16
ELEMENTARY SIMPLE	4	m. 6 f. 8 14	m. 1 f. 0 1
BONY MOTOR (LEPTOSOME)	10	m. 6 f. 5 11	m. 3 f. 9 12
FLESHY MOTOR	20	m. 12 f. 13 25	m. 3 f. 3 6
SENSITIVE SMALL	18	m. 5 f. 7 12	m. 2 f. 14 16
SENSITIVE LONG	12	m. 1 f. 3 4	m. 4 f. 4 8
ACROMICRY	4	f. 2	f. 7

TABLE IIA

deals with the Numerical Incidence of Certain Characteristics of the Size of the Fingers in Relation to the Different Hand-Types in 80 Paraphrenic Patients

HAND-TYPE	DIGITAL FORMULA		FIFTH FINGER TOO LONG	FIFTH FINGER TOO SHORT	EQUAL LENGTH OF THIRD + FOURTH FINGERS	THUMB TOO LONG	THUMB TOO SHORT	ANKYLOSIS OF FIFTH FINGER	ANKYLOSIS OF ALL FINGERS	INDEX TOO SHORT
	HUMAN	SIMIAN								
ELEMENTARY REGRESSIVE LONG FINGERS	2	2	1	2	0	2	0	1	0	0
ELEMENTARY REGRESSIVE SHORT FINGERS	0	8	0	3	0	0	1	1	0	1
ELEMENTARY SIMPLE	0	4	2	1	0	0	2	0	0	0
BONY MOTOR (LEPTOSOME)	6	4	2	1	3	1	1	0	0	0
FLESHY MOTOR	5	15	2	3	0	1	8	0	0	1
SENSITIVE SMALL	4	14	3	9	2	0	8	0	0	1
SENSITIVE LONG	7	5	4	4	2	2	2	0	0	0
ACROMICRY	0	4	0	1	1	0	2	0	0	1

TABLE IIB

deals with the Numerical Incidence of Certain Characteristics of the Size of the Fingers in Relation to the Different Hand-Types in 80 Manic-Depressive Patients

HAND-TYPE	DIGITAL FORMULA		FIFTH FINGER TOO LONG	FIFTH FINGER TOO SHORT	EQUAL LENGTH OF THIRD + FOURTH FINGERS	THUMB TOO LONG	THUMB TOO SHORT	ANKYLOSIS OF FIFTH FINGER	ANKYLOSIS OF ALL FINGERS	INDEX TOO SHORT
	HUMAN	SIMIAN								
ELEMENTARY REGRESSIVE LONG FINGERS	1	5	1	3	0	1	0	1	0	1
ELEMENTARY REGRESSIVE SHORT FINGERS	1	5	1	3	0	0	3	1	0	0
ELEMENTARY SIMPLE		14	1	0	1	0	0	1	0	0
BONY MOTOR (LEPTOSOME)	5	6	4	0	0	0	0	0	0	1
FLESHY MOTOR		25	0	0	1	0	3	0	0	2
SENSITIVE SMALL		12	0	4	0	1	1	0	0	0
SENSITIVE LONG		4	0	1	0	0	0	0	0	2
ACROMICRY		2	0	0	0	0	1	0	0	1

TABLE IIc

deals with the Numerical Incidence of Certain Characteristics of the Size of the Fingers in Relation to the Different Hand-Types in 80 Epileptic Patients

HAND-TYPE	*DIGITAL FORMULA*		FIFTH FINGER TOO LONG	FIFTH FINGER TOO SHORT	EQUAL LENGTH OF THIRD + FOURTH FINGERS	THUMB TOO LONG	THUMB TOO SHORT	ANKYLOSIS OF FIFTH FINGER	ANKYLOSIS OF ALL FINGERS	INDEX TOO SHORT
	HUMAN	SIMIAN								
ELEMENTARY REGRESSIVE LONG FINGERS	3	13	5	0	4	1	3	1	0	5
ELEMENTARY REGRESSIVE SHORT FINGERS	1	13	3	3	1	2	7	2	0	9
ELEMENTARY SIMPLE	0	1	1	0	0	1	0	0	0	0
BONY MOTOR	2	10	3	2	0	1	0	1	1	1
FLESHY MOTOR	1	5	1	0	1	1	0	0	0	1
SENSITIVE SMALL	6	10	2	4	1	2	3	1	0	3
SENSITIVE LONG	4	4	2	0	1	0	1	2	0	1
ACROMICRY	1	6	0	2	2	0	2	0	0	1

TABLE IIIA

deals with the Numerical Incidence of Certain Characteristics of the Fingers in 80 Paraphrenics, 80 Manic-Depressive Patients and 80 Epileptics
(m. = male; f. = female)

MENTAL ILLNESS	*FINGER-TIPS*				*NAILS*				*MISSHAPEN*	
	SPATULATE	CONICAL	SQUARE	HYPER-FLEXIBLE	LONG	MEDIUM	SMALL	UNDER-DEVELOPED	FIFTH FINGER	THUMB
PARAPHRENICS	15	23	42	13	9	21	28	22	15	18
MANIC-DEPRESSIVE PATIENTS										
m.	10	5	20	1	3	6	20	6	5	6
f.	0	13	32	5	5	16	18	6	15	7
	10	18	52	6	8	22	38	12	15	13
EPILEPTICS										
m.	1	18	11	5	5	8	9	9	6	11
f.	15	15	20	7	3	14	15	17	17	17
	16	33	31	12	8	22	24	26	23	28

TABLE IVA

deals with the Numerical Incidence of Abnormalities of the Creases and Papillary Ridges in 80 Paraphrenic, 80 Manic-depressive and 80 Epileptic Patients

MENTAL ILLNESS	SIMIAN LINE	LONG LONGI-TUDINAL LINE MISSING	CONFUSED CREASE-LINE PATTERN	FAULTS UPPER TRANSVERSE LINE	FAULTS LOWER TRANSVERSE LINE	WHORLS OR LOOPS THENAR EMINENCE	WHORLS OR LOOPS HYPO-THENAR EMINENCE	ARC-TRIANGLE PATTERN ONE OR TWO FINGERS	ARC-TRIANGLE PATTERN ALL FINGERS
PARAPHRENIA	9	0	3	43	48	2	25	23	5
MANIC-DEPRESSIVE PSYCHOSIS	4	0	0	42	55	5	22	15	1
EPILEPSY	12	3	4	45	53	5	18	33	8

ideas of reference to violent feelings and thoughts of being deprived, persecuted and done in by 'enemies'. I have described in full the workings of the mind of some paraphrenics in my book *A Psychology of Gesture*, and shall not go into details here. It suffices to say that paraphrenic patients may appear to be quite normal most of the time, until their condition reveals itself, either through sudden aberrations in argument, flight into ideas of reference or imaginations of persecution. Their facial expression and gestures as well as postures alter according to the contents of thought and the pressure of emotions.

DISCUSSION OF THE TABLES

My material refers to the hands of 240 patients, 80 in each of the three groups: paraphrenia, manic-depressive psychosis and epilepsy. In the first group all patients were women, in the second 45 were women and 35 men and in the third 50 were women and 30 men.

The material has been listed in 6 tables and the same hand-features have been investigated as in schizophrenics.

In order to relate these tables to the previous ones dealing with hand-characteristics in schizophrenia, I refer, for example, to Table IA or IIA in order to show that the same items are listed as in the corresponding Tables I and II of the previous study.

DISCUSSION OF TABLE IA

I have compared the types of hands found in 80 paraphrenics with those found in 80 manic-depressive patients and in 80 epileptics.

The most interesting result of this investigation seems to me to be that in both conditions, paraphrenia and manic-depressive psychosis, the fleshy motor type of hand is most frequently present: it occurred in 20 paraphrenics and in 25 manic-depressive patients. I must emphasize this finding in view of its differential diagnostic value. I observed the same hand-type in only 15 out of 245 schizophrenics, and one may assume that the broad and fleshy hand with short and strong fingers belongs to what Kretschmer has termed the 'cycloid' temperament.

Now, the only query about such an interpretation is the fact that this hand-type is common to both types of mental illness, paraphrenia and manic-depressive psychosis, and I have said in the beginning of this chapter that the paraphrenic patient resembles more the schizophrenic in his abnormal reactions than the manic-depressive personality.

I should perhaps have mentioned at once that paraphrenia contains elements of both the principal psychoses, by which I mean schizophrenia and manic-depressive psychosis.

This statement is confirmed by another result of this table: namely the predominance of the sensitive small and long hands in paraphrenics, where they number 18 and 12 respectively, compared with 12 and 4 in manic-depressive patients.

The fact that we find the most 'normal' hands, fully developed, muscular and broad, what may be called active hands among manic-depressive patients, can be taken to mean that this type of psychosis attacks people who are by constitution more fully developed than schizophrenics, whose hands so often reveal an over-development of the factors of sensitivity and imagination, together with infantilism physical and emotional.

The fact that we find a great number of active hands, but an even greater number of 'sensitive' hands particularly of the small type in paraphrenics may be taken to mean that paraphrenia occupies an intermediate place between the two other psychoses. While paraphrenics are more withdrawn into purely imaginative complexes than manic-depressive patients, they are still nearer to a normal reaction pattern than schizophrenics.

This situation is also borne out by the fact that the elementary simple hand is found in 14 cases of manic-depressive psychosis but in only 4 of paraphrenia, and that there were 4 cases of acromicry among paraphrenics but only 2 in manic-depressive patients.

The elementary regressive hand occurred most often in the epileptic patients investigated. It was present in 18 of the 30 male and in 12 of the 50 female patients. The sensitive small hand takes second place. 16 out of 80 epileptics possessed this type of hand, 14 women and 2 men. This marked difference in the sexes shows (and I have pointed this out before) that the sensitive small hand is really a woman's hand.

The bony motor and fleshy motor hands with 12 and 6 cases respectively have third and fourth place in the scale of frequency.

The greatest number of dysplastic hands were observed among epileptics, namely in 7 cases. One pair of these hands showed the lack of the interdigital fold between the terminal and middle phalanxes of the fifth digit,* a hand-trait that had been previously attributed to mongolian imbeciles only.

This table shows that the elementary regressive hand is the predominant one among epileptics, as it is with mental defectives. The condition of epilepsy can obviously not be traced from hand-features, but only the constitutional type that generally goes with it. It is therefore not surprising that the hands of epileptics show similar marks to those of mental defective- and certain schizophrenic patients.

TABLES IIA, IIB AND IIC

These tables refer to the numerical incidence of characteristics of the size of the fingers in relation to different hand-types in

Table IIA—80 female paraphrenic patients
Table IIB—80 manic-depressive patients
Table IIC—80 epileptic patients.

THE DIGITAL FORMULA

In all three groups investigated the simian digital formula is more frequently found than the human one, but the latter is far more often present in the hands of mental patients than in normal individuals. It occurred in 24 paraphrenic, in 18 epileptic and in 7 manic-depressive patients, another proof for the fact, which I have already mentioned, that the hands of the last group resemble most closely those of people without mental illness. Paraphrenics are closest to schizophrenics in regard to this hand-trait.

THE FIFTH FINGER

An Overlong Fifth Finger was, as I have shown, significantly frequent in schizophrenic patients. This hand-feature occurred 14 times in the hands of paraphrenics, 7 times in those of manic-depressive patients and 17 times in those of epileptics. It is most rarely found in the elementary regressive and fleshy motor hands, and occurred most often in the sensitive long type of hand.

A Too Short Fifth Finger was most frequently present in the hands of paraphrenics, where it occurred in 24 cases. In either of the other two groups it occurred 11 times. I have already pointed out that this hand-trait has an endocrine significance and is of no special diagnostic significance for mental illness as such.

Equal or Almost Equal Length of the Medius and Fourth Finger. This hand-trait represents an interesting abnormality of the digital formula and, as I have shown, is significantly often present in schizophrenics. It occurred 8 times in paraphrenics, twice in manic-depressive patients and 10 times in epileptics.

THE THUMB

According to my experience, a thumb that is too short is far more frequent than the opposite both in the hands of 'ordinary' and abnormal people. We observe again, as in other abnormalities of the fingers, that too short a thumb is found least often in the hands of manic-depressive patients, most often (in 23 cases) in paraphrenics, and quite frequently (in 16 cases) in epileptics.

A thumb that is too long occurred only twice in manic-depressive patients, 6 times in paraphrenic and 8 times in epileptic patients.

ANKYLOSIS

Ankylosis of the fifth finger is a well-known mark of degeneracy, as I have already pointed out. It occurred most often in epileptics (in 7 cases), 3 times in manic-depressive and only twice in paraphrenic patients. Ankylosis of all fingers is a very rare abnormality and is generally found in mental defectives only. It was present only in one epileptic patient.

DISCUSSION OF TABLE IIIA

This table deals with the shape of the fingers and nails, the flexibility of the finger-tips and certain abnormalities in shape of the thumb and fifth finger.

THE FINGER-TIPS

The shape of the finger-tips goes, on the whole, with the general shape of the hand. Its length and either square, conical or spatulate form are, in the first place, conditioned by the endocrine constitution. I have already pointed out that short and spatulate finger-tips are characteristic of hypothyroic functioning, the long, conically shaped terminal phalanx, which is mostly found in the sensitive long hand, is characteristic of subgonadic functioning. One can consider the well-proportioned, square finger-tip to be the most 'normal' one. Indeed, this type of finger-tip is most frequent in the hands of normal people. The spatulate finger-tip is, according to my experience, rarest in 'normal' persons and is comparatively often found in mental patients.

The spatulate finger is more frequently found than any other in schizophrenics and can be considered an important characteristic of the schizophrenic hand, as I have mentioned earlier in this chapter. Among the other types of mental patients it occurred most often in epileptics (in 16 cases) and was least frequently found in manic-depressive patients. Tapered finger-tips were present in 33 epileptic patients and in 23 paraphrenic women, and occurred in 18 manic-depressive patients.

In regard to the square finger-tip the manic-depressive patients had first place.

FLEXIBILITY OF THE FINGER-TIPS

Hyperflexible finger-tips, the significance of which I have already discussed, occurred less often in these three groups of patients than in the schizophrenics investigated.

The manic-depressive patients showed this hand-trait in only 6 cases, and the two other groups in 13 cases of paraphrenia and in 12 of epilepsy.

ABNORMAL SHAPE OF THE THUMB AND FIFTH FINGER

Apart from abnormality in size, one finds hands with misshapen marginal digits, of which the 'clubbed' thumb, called the 'murderer's' thumb in chiromancy, is a well-known example. Such abnormalities in shape go mostly with the elementary regressive type of hand. Both the misshapen fifth finger and thumb are most frequently found in the epileptic group of this study. A deformed thumb occurred in 28 epileptics compared with only 18 paraphrenics and 13 patients with manic-depressive psychosis. A misshapen fifth finger was equally frequent in paraphrenics and patients with manic-depressive psychosis, namely in 15 cases of each, and occurred in 23 epileptics.

THE NAILS

The length of the nail is, as a rule, dependent on the length of the hand and the fingers in particular. A long nail goes generally with a long hand and long fingers, a short one with a small hand and short and stout fingers. But this rule does not always hold good as one observes often enough short and even rudimentary nails in elementary hands with long fingers. We may perhaps make endocrine dysfunctioning responsible for such a deviation from a structural tendency.

The Long Nail was the least frequently found in all patients of the three groups. It was most often present in the group of paraphrenics, and occurred in either of the other two in only 8 cases.

The Under-developed nail was most often present in the hands of epileptic patients. This nail-form occurred in only 12 cases in manic-depressive individuals, compared with 26 epileptics and 22 paraphrenics.

The Small Nail was characteristic of the hands of manic-depressive patients, where it occurred 38 times, in almost 50 per cent, while I observed it only 24 times in epileptics and 28 times in paraphrenics.

DISCUSSION OF TABLE IVA

ABNORMALITIES OF THE CREASES AND PAPILLARY RIDGES

The group of epileptics heads the list in regard to abnormalities of the creases except for the frequency of faults in the Lower Transverse line.

Both the presence of the simian- and the absence of the long-longitudinal line are found significantly often in the epileptic group. Both abnormalities are, as I have mentioned on several occasions in this book, characteristic of the elementary regressive type of hand. The absence of the long longitudinal line is, naturally, more often seen in this type of hand with short fingers.

While the two items just mentioned correspond to what one would expect, the fact that faults in the Lower Transverse line are most frequent in the hands of manic-depressive patients, is a surprise. Faults in creases are, as I also mentioned before, common in both the hand of the normal and the abnormal individual, though they are more pronounced in quality and quantity in the latter. This hand-feature is, however, of minor diagnostic significance.

ABNORMALITIES OF THE PAPILLARY RIDGES

Here again one finds that not much difference exists in the frequency of an unusual location of certain patterns of the papillary ridges in the three groups.

In all three types of patients such abnormalities were found comparatively often, particularly the whorls and loops on the hypothenar eminence.

I must mention that the arc-triangle pattern on one or two finger-tips occurred most often in the hands of epileptic patients, where I observed it 33 times, compared with 23 times in paraphrenics, and only 15 times in manic-depressive patients.

The same holds good for the occurrence of the arc-triangle pattern on all finger-tips, which is very rarely found in the hands of normal people. The hand-trait concerned was found in 8 epileptics compared with one manic-depressive patient and 5 paraphrenics.

At the end of this chapter I want to summarize the most important differences which I have found in the hands of schizophrenics compared with those of the three other groups of mental patients.

1.

The 'active' or motor-type of hand was rarely found in schizophrenics; the elementary simple hand was practically absent in 245 schizophrenic patients.

The elementary regressive hand was found comparatively often, particularly in catatonic schizophrenia and schizophrenia simplex; these are the most severe forms of schizophrenia.

The 'passive' or sensitive type of hand was observed significantly often in schizophrenia and particularly in the paranoid form.

2.

In comparing the frequency of 'active' with 'passive' hands, the manic-depressive patients possessed the greatest number of the first: namely 14 times elementary simple and 25 times fleshy motor hands.

The sensitive small hand occurred far more often than the sensitive long one among manic-depressive patients, a fact that emphasizes the greater 'normality' of hand-forms in these patients, as this variety of the sensitive type of hand is the more 'active' one and belongs to more adaptable personalities than the sensitive long hand.

3.

One of the most interesting observations of this study is the fact that paraphrenic patients show such a high percentage of 'active' hands. Though their number lags behind that of the previous group, it is markedly greater than with schizophrenic patients.

One may therefore believe that the paraphrenic patient manifests traits of both: schizophrenia and manic-depressive psychosis. According to my observations, it seems likely that the imaginative processes resemble those characteristic of schizophrenia and the emotional reactions those typical of manic-depressive psychosis, though they appear to be less violent.

As to the detailed observations on fingers and nails, the hands of schizophrenics showed marked differences to those of the other three groups of patients.

4.

The spatulate finger-tip is much more often found in schizophrenics than in other mental patients; and so is the overlong finger-tip.

While the spatulate finger-tips are predominant and the tapered ones very frequent, the square finger-tip is relatively rare in the hands of schizophrenic patients. Square finger-tips are, on the other hand, most frequently found in manic-depressive psychosis and paraphrenia.

5.

The long nail as well as that of medium size is usual in the hands of schizophrenics, while the situation is quite different in the other three groups. The small nail, both in its normal and in its rudimentary form, are predominant in their hands.

6.

Hyperflexibility of the finger-tips seems to be a marked characteristic of paranoid schizophrenia—but not for the other forms of this psychosis.

7.

The misshapen thumb and fifth finger are less frequent in schizophrenics than in the other mental patients of this study, excepting only patients with secondary dementia. One finds among the latter hand-traits that resemble those present in congenital mental defectives.

8.

In regard to gross abnormalities of the crease-lines of the palm there is not much difference in the different psychoses. It is however noteworthy that faults of both the upper and lower transverse lines are less frequent in the hands of schizophrenics than in those of other mental patients and the same applies to the occurrence of whorls and loops on the thenar and hypothenar eminences. While I consider it to be justified to speak of certain diagnostic marks in the hands of schizophrenic patients, I must make it quite clear that one cannot classify hands as 'schizophrenic', 'manic-depressive' or 'paraphrenic' hands. The hand-traits in schizophrenic psychosis seem to be more definite than those in any other mental illness, obviously because extreme abnormalities manifest themselves more clearly in hand-traits than do less striking ones. From a diagnostic point of view, therefore, the hand of the schizophrenic will act as a guide to the psychiatrist, and so will the hand of the mental defective, but that of the epileptic will only show marks of mental deficiency.

The hands in manic-depressive psychosis and paraphrenia are more interesting for the possibility they offer of assessing remnants of integrating qualities and other characteristics of personality than for psychiatric diagnosis.

To say it once more, the hand can only show constitutional tendencies of the psyche-soma, and the indications which it can give must be followed up and verified by other diagnostic methods.

CHAPTER VIII

HANDS OF CHILDREN

A Comparative Study of Hands of Normal and Mentally Defective Children

GENERAL ASPECT

IN the previous chapters on my clinical studies the material has been, as the term implies, exclusively pathological. In investigating the hands of abnormal people I had two ideas in mind:

1. To learn about those traits of the hand which are relevant for the diagnosis of psychological and mental illness; and
2. To find out how the experience and knowledge I had gained from studying the abnormal may be applied to the normal personality.

I have repeatedly emphasized in this book and in earlier writings that the frontier line between the healthy psycho-physical organism and the sick one is but narrow and that there are many paths which connect the lands of the sane and the insane.

This must be kept in mind while one compares possibilities of diagnosis in regard to the defective and to the fully developed personality. I prefer this expression to the word 'normal' as a normal person only exists (as I need hardly mention) in phantasy and not in reality. I have left to the last in these clinical studies and to the end of this book the application of my method of diagnosis to the hands of normally developed persons and for this I have chosen the adolescent, for reasons both theoretical and practical. For simplification and clarity my diagnostic estimates have been focused first of all on one particular aspect. This subject is new in psychological research and it seemed to me essential that it should seek information about one aspect of personality more tangible than the emotions—namely intelligence. There was then good reason for comparing the hands of normally intelligent children with those of mental defectives, to use a term which is misleading, for, as I have already stressed, it is the whole personality and not merely the mental equipment which is defective or insufficient in these children. This must not be forgotten, but the point of comparison here to be considered is the difference in intellectual capacity.

Apart from theoretical reasons my choice was dictated by practical considerations. First of all, I had been studying the hands of mental defectives, both children and adults, in a particularly large and significant number, and so had gained much experience of the physiognomy and physical traits of the hand connected with this condition. It seemed thus obvious that I should make use of this in a statistical comparative study of normal and abnormal cases. Secondly, I thought the possibility of an easy and reliable diagnostic estimate as to normal or defective intelligence of very great importance for reasons I also have mentioned earlier in this book, namely the possible misplacement of a normally intelligent child among subnormal children. The harm that can be done to such an individual is untold and terrifying. It need hardly be pointed out as it is so obvious. Thirdly, but of no less importance, to my mind, was the consideration that in any study of human beings that of the adolescent deserves a particular place, as it entails the study of potentialities far more poignantly than at any other phase of human existence. In the adolescent one finds an accumulation of constructive and destructive capacities and it is essential to help the growing child at the right time with the right guidance and the right freedom. The value of psychological studies of the adolescent mentality is not only an asset for the education and development of personality but it is also of important practical value to establish a diagnostic estimate of mentality and temperament, in order to be able to give vocational guidance safely to the young and to save mistakes and waste of time at the outset of some specific vocational training. It is a banality to say that talents are not by themselves sufficient to serve as indicators for a future profession; temperament and character must fit in with the demands of any profession and both vocation and personality must be considered carefully.

Here then were my reasons for comparing the hands of 120 normally intelligent and 120 mentally defective children with a view to their possible significance for diagnosis.

I have chosen to compare the hands of boys and girls in equal numbers. A certain number of mentally defective boys of this study was not of the group whose hands I described in the chapter 'The Hand of the Mental Defective'.

The age of the children ranged from 14 to 16 years. Fourteen of the normal girls were just under 14 years of age; they were particularly intelligent children who were in the same form with the older ones.

I was given the opportunity of studying the hands of children at several London schools and I am very grateful for the interest and help that I found there among both teachers and pupils.

The statistical results of this study are listed in four columns showing certain measurements in normal and mentally defective children and in three tables showing the numerical incidence of characteristic hand-traits on lines similar to those used in the previous investigation on mental patients.

In order to give the reader an idea of the kind of diagnosis which can be made in a minimum of time from hand-features and also in order to show how these diagnostic estimates compare with others, I have divided the contents of this chapter into two parts; the first one dealing with the general aspect of hand-diagnosis on account of the statistical results and the second one dealing with the individual aspect of hand-diagnosis. The second part contains a comparison between:

1. Hand diagnosis and psychiatric diagnosis in 25 cases of high-grade and difficult boys
2. Hand-diagnosis and estimates of several teachers on 19 normally intelligent girls.

My diagnostic estimates on the mentally defective boys are very short and only refer to essential characteristics of mentality and temperament. My statements on the 19 girls are more complex and contain more detailed observations.

Boys and girls of whose hands I have made these more detailed interpretations were mostly of the same age-groups as those whose hands I had investigated for the general and statistical study of this chapter.

I have treated the material at my disposal in the simplest possible way in order to enable the reader to recognize at a glance what the essential differences are between the hands of normal and mentally defective children. I thought it important for the more complete understanding of the constitutional make-up of the children concerned to list certain body-measurements in all cases: namely height and weight of each child together with the length and width of the hands.

I have differentiated two kinds of width:

W1 representing the width of the palm measured just underneath the fingers, w2 representing the broadest part of the palm, which, as a rule, is represented by the distance between the radial and ulnar edge measured just above the insertion of the thumb.

All the mentally defective children investigated in this study are high-grade, which means that their I.Q. lies between 60 and 100. Six boys and 8 girls of this group are of almost normal intelligence, having an I.Q. of about 90 or over 90.

It seemed desirable to compare the hands of children in whom the difference of intellectual capacity is but mildly, though significantly, marked, in order to find out whether there are definite constitutional stigmata in the hands of the feebleminded. These could, of course, not be as dramatic as those found in the lower grades of imbecility but as I am able to show they are sufficiently obvious to provide some general guide towards diagnosis.

In a few cases of boys with mental deficiency I made the diagnosis of normal intelligence, to which the psychiatrist agreed. The I.Q., one must remember, often expresses mental backwardness where there is no mental deficiency, but where emotional disturbance produces mental achievements which are in no way proportionate to the intellectual capacity.

Most of the high-grade boys and girls I have dealt with in this investigation had come into trouble with the police for more or less minor offences. They are problem children, some of them with openly destructive tendencies.

The question arises if these character disturbances are mainly the outcome of educational mismanagement in an unfortunate social milieu or if there is a definite constitutional disposition towards delinquency. We must call a tendency towards delinquency the existence of persistent destructive attitudes which for one reason or another cannot be controlled by the child. One can therefore speak of moral inferiority as we can speak of mental inferiority. I do not mean the natural destructiveness that goes with the evolutional development of the child, of which every parent and teacher should know. Present-day psychology has a very strong leaning to ascribe emotional disturbances in the child mainly to environmental factors and to overlook or to underrate the constitutional predisposition of the child. Nobody can deny that the background is of the greatest importance for the development of character and that a bad and asocial family can spoil the good dispositions of a child for ever. As a great number of the high-grade children with character difficulties came from broken or otherwise bad homes, it is obvious that the environmental factor played a great part in their delinquent tendencies. The case histories of these children reveal so often emotional conflicts far too difficult to be borne without permanent damage to the personality that it is clear that the social frame, represented by the family, and the cultural pattern, represented by the community, must have a great and sometimes overpowering influence on the young.

This side of the situation has been most eloquently and convincingly described by D. H. Stott in his book *Delinquency and Human Nature*. In spite of such a well-founded investigation as Stott has made, who puts early delinquency down to emotional conflicts in bad or asocial surroundings, I am of the opinion that the constitutional factor of delinquency has not been sufficiently recognized by him and many other protagonists of modern child psychology. Apart from Sheldon, no major attempt has been made in the last twenty years to discover the constitutional dispositions of man.

Any direction of research entails a subconscious bias. When one looks for the environmental factors shaping personality one is bound to overlook much of the constitutional ones and vice versa. I think that few research workers, if at all, can escape this pitfall. And I would never claim to be an exception to this rule. I have naturally looked at and looked for the constitutional factor in personality as seen from hand-features and allowing for a bias towards it, I am convinced of its existence and its immense importance as an indication of the origin of much asocial behaviour. It is for me to show by the results of this investigation that a difference does exist between the physical traits, particularly those of the hand, shown by the normally developed and by the defective child.

HANDS OF NORMAL BOYS

CASES	AGE	HEIGHT (CM.)	WEIGHT (KG.)	LENGTH OF HAND (CM.)	WIDTH OF HAND W1 (CM.)	WIDTH OF HAND W2 (CM.)
1	13·4	155	45·4	R 18·8	R 7·3	R 7·9
				L 18·7	L 7·0	L 7·6
2	14	159	48·1	R 18·3	R 7·6	R 8·5
				L 18·1	L 7·3	L 7·9
3	14	169	56·8	R 20·2	R 9·0	R 8·5
				L 20·3	L 9·4	L 9·0
4	14	151·4	44	R 17·5	R 7·5	R 8·0
				L 17·9	L 7·4	L 7·8
5	14·2	169	50·8	R 19·2	R 7·7	R 8·0
				L 18·2	L 7·5	L 7·8
6	14·3	164	38·2	R 17·4	R 7·5	R 7·8
				L 17·7	L 7·4	L 7·8
7	14·4	145	34·5	R 17·0	R 7·1	R 7·5
				L 17·3	L 6·9	L 7·3
8	14·4	160	47·6	R 17·9	R 7·7	R 8·4
				L 18·2	L 7·2	L 7·7
9	14·4	162·7	44·4	R 18·9	R 7·6	R 8·1
				L 19·0	L 7·5	L 8·1
10	14·4	162·7	47·6	R 18·7	R 7·7	R 8·7
				L 18·6	L 7·3	L 7·6
11	14·4	161·4	50·8	R 19·3	R 8·1	R 8·7
				L 19·3	L 8·2	L 8·9
12	14·4	175·2	60·4	R 20·0	R 8·5	R 9·0
				L 20·5	L 8·5	L 9·0
13	14·4	160	46·2	R 17·8	R 8·0	R 8·2
				L 18·2	L 7·5	L 7·9
14	14·5	157·6	50	R 18·2	R 7·8	R 8·1
				L 18·8	L 7·5	L 8·1
15	14·5	156·4	44·4	R 17·5	R 7·2	R 7·8
				L 17·5	L 7·1	L 7·6
16	14·5	145	35·8	R 16·7	R 6·8	R 7·0
				L 16·5	L 6·5	L 7·0
17	14·5	160	47·2	R 18·5	R 8·0	R 8·4
				L 18·8	L 7·9	L 8·2
18	14·5	159	59·5	R 18·2	R 7·9	R 8·4
				L 18·4	L 7·8	L 8·0
19	14·6	157·6	48·6	R 18·7	R 8·5	R 9·0
				L 19·1	L 8·0	L 8·7
20	14·6	160	40·4	R 18·2	R 7·2	R 7·7
				L 18·5	L 7·6	L 7·7
21	14·6	166·6	46·3	R 19·0	R 7·9	R 8·4
				L 19·4	L 7·5	L 8·2

CASES	AGE	HEIGHT (CM.)	WEIGHT (KG.)	LENGTH OF HAND (CM.)	WIDTH OF HAND W1 (CM.)	WIDTH OF HAND W2 (CM.)
22	14·6	156·4	41·3	R 18·2	R 7·5	R 8·2
				L 18·3	L 7·1	L 7·6
23	14·7	157·6	49	R 17·6	R 7·8	R 8·3
				L 17·8	L 7·4	L 8·2
24	14·7	170·3	55·4	R 20·4	R 8·7	R 9·1
				L 20·5	L 8·6	L 8·7
25	14·7	175·2	59·5	R 21·0	R 8·8	R 9·1
				L 21·4	L 8·1	L 8·6
26	14·7	165·3	51·8	R 19·5	R 8·0	R 8·3
				L 19·8	L 8·2	L 8·2
27	14·7	161·4	52·65	R 18·2	R 8·2	R 8·6
				L 18·2	L 8·2	L 8·4
28	14·7	161·4	40·4	R 17·8	R 8·1	R 8·3
				L 18·2	L 7·8	L 8·0
29	14·9	156·4	41·3	R 17·6	R 7·4	R 8·2
				L 18·0	L 7·0	L 7·6
30	14·9	151·4	41·3	R *16·5*	R 7·8	R 8·1
				L *16·8*	L 7·5	L 8·0
31	14·9	164	51·8	R 18·5	R 7·9	R 8·2
				L 18·6	L 7·8	L 8·1
32	14·9	152·6	36·25	R 17·0	R 7·0	R 7·2
				L 17·0	L 6·8	L 7·0
33	14·10	180·2	63·2	R 20·0	R 8·7	R 9·2
				L 20·2	L 8·4	L 9·0
34	14·10	166·6	46	R 19·6	R 7·6	R 8·5
				L 19·8	L 7·8	L 8·4
35	14·11	151·4	39·9	R 17·5	R 7·7	R 8·2
				L 17·5	L 8·0	L 8·2
36	14·11	159	41·3	R 18·8	R 8·2	R 8·0
				L 18·8	L 8·0	L 7·5
37	15	172·7	54	R 19·1	R 8·2	R 8·9
				L 19·2	L 8·2	L 8·9
38	15	172·7	57·7	R 20·1	R 8·5	R 9·5
				L 20·3	L 8·3	L 9·1
39	15	167·8	61·3	R 19·5	R 8·0	R 9·0
				L 19·4	L 8·1	L 8·8
40	15	155	53·6	R 19·1	R 8·5	R 9·0
				L 18·6	L 7·9	L 8·6
41	15	164	75·8	R 18·2	R 8·5	R 9·0
				L 18·8	L 8·3	L 9·2
42	15·1	165·3	51·6	R 20·3	R 8·5	R 9·1
				L 20·1	L 8·1	L 8·4
43	15·1	141·2	32·2	R 15·6	R 7·0	R 7·3
				L *16·0*	L 6·7	L 7·1
44	15·1	164	44·4	R 18·5	R 8·2	R 8·6
				L 18·7	L 8·0	L 8·6

CASES	AGE	HEIGHT (CM.)	WEIGHT (KG.)	LENGTH OF HAND (CM.)	WIDTH OF HAND W1 (CM.)	WIDTH OF HAND W2 (CM.)
45	15·1	171·5	50·8	R 19·8	R 7·9	R 8·8
				L 20·0	L 7·8	L 8·1
46	15·4	164	49·4	R 19·2	R 8·0	R 8·5
				L 19·6	L 8·0	L 8·4
47	15·4	160	49	R 18·6	R 7·8	R 8·2
				L 18·6	L 7·4	L 8·0
48	15·5	161·4	48·6	R 17·6	R 7·5	R 8·2
				L 18·0	L 7·2	L 7·6
49	15·6	164	44·4	R 18·1	R 7·6	R 8·2
				L 18·3	L 7·6	L 8·0
50	15·6	174	63·5	R 19·0	R 8·1	R 9·1
				L 19·3	L 8·1	L 8·8
51	15·6	179	68	R 20·6	R 9·0	R 9·5
				L 21·0	L 9·0	L 9·7
52	15·7	161·4	50·4	R 19·2	R 8·0	R 8·8
				L 19·5	L 8·4	L 9·0
53	15·8	172·7	54·9	R 18·5	R 7·9	R 8·3
				L 18·9	L 7·8	L 8·3
54	15·9	169	50·8	R 19·5	R 7·9	R 8·8
				L 19·4	L 7·8	L 8·3
55	15·10	175·2	64·8	R 19·3	R 8·4	R 9·5
				L 19·6	L 8·0	L 8·7
56	15·11	171·6	51·8	R 18·7	R 8·0	R 8·9
				L 19·0	L 8·0	L 8·6
57	15·11	167·8	49	R 18·7	R 8·5	R 9·2
				L 18·5	L 8·2	L 8·8
58	16·0	159	42·2	R 18·0	R 8·0	R 8·5
				L 18·1	L 7·8	L 8·2
59	16·6	162·7	51·7	R 18·5	R 8·3	R 9·0
				L 19·0	L 8·2	L 9·2

Note: I have put into italic figures the measurements of abnormally small hands which represent cases of acromicry.

Note: In one case of this group of boys the measurements of height and weight were not available.

HANDS OF MENTALLY DEFECTIVE BOYS

CASES	AGE	HEIGHT (CM.)	WEIGHT (KG.)	LENGTH OF HAND (CM.)	WIDTH OF HAND W1 (CM.)	WIDTH OF HAND W2 (CM.)
1	13·8	153·8	43·6	R 17·6	R 7·5	R 8·2
				L 17·2	L 7·5	L 7·8
2	14	143·8	31·3	R 17·2	R 7·0	R 7·6
				L 17·6	L 7·1	L 7·5
3	14	152·6	38·1	R 17·5	R 7·0	R 8·3
				L 18·0	L 7·0	L 7·9
4	14	136·1	35·8	R 16·9	R 7·3	R 8·3
				L 16·9	L 7·2	L 7·9
5	14	141·2	38·8	R *16·3*	R 7·0	R 7·5
				L *16·6*	L 7·0	L 7·5
6	14	155	43·1	R 17·2	R 7·5	R 8·0
				L 17·4	L 7·2	L 7·8
7	14	148·8	35·8	R 18·1	R 7·3	R 7·7
				L 18·8	L 7·2	L 7·6
8	14	148·8	42·25	R 17·3	R 7·2	R 7·6
				L 17·5	L 7·3	L 7·6
9	14	148·8	38·8	R 16·9	R 7·1	R 7·8
				L 17·1	L 7·2	L 7·6
10	14	150	42·6	R 18·2	R 7·8	R 8·0
				L 18·5	L 7·5	L 7·8
11	14	160	46·1	R 17·5	R 7·4	R 8·0
				L 17·9	L 7·4	L 7·6
12	14	156·4	42·9	R 18·3	R 7·6	R 7·8
				L 18·6	L 7·5	L 7·7
13	14	141·2	37·6	R *16·5*	R 6·8	R 7·5
				L *16·6*	L 7·0	L 7·3
14	14·1	142·5	39·5	R 17·2	R 7·5	R 8·0
				L 17·3	L 7·5	L 8·3
15	14·1	156·4	49·35	R 18·5	R 8·0	R 9·6
				L 18·5	L 7·8	L 8·5
16	14·1	152·6	52	R 18·5	R 8·0	R 8·6
				L 18·5	L 8·0	L 8·1
17	14·2	147·6	37·9	R 17·5	R 7·0	R 7·8
				L 17·8	L 7·3	L 7·6
18	14·2	134·8	28·3	R *16·1*	R 6·3	R 6·8
				L *16·5*	L 6·2	L 6·6
19	14·2	142·5	31·3	R 17·0	R 6·5	R 6·8
				L 17·0	L 6·5	L 6·7
20	14·2	147·6	44·3	R 17·1	R 7·5	R 8·0
				L 17·5	L 7·5	L 8·0
21	14·3	145	37·6	R 17·6	R 7·1	R 7·5
				L 17·4	L 7·6	L 7·2

CASES	AGE	HEIGHT (CM.)	WEIGHT (KG.)	LENGTH OF HAND (CM.)	WIDTH OF HAND W1 (CM.)	WIDTH OF HAND W2 (CM.)
22	14·3	140	33·1	R 16·8	R 7·1	R 7·6
				L 16·6	L 6·5	L 6·7
23	14·3	166·6	50·8	R 19·0	R 8·0	R 8·3
				L 19·4	L 7·8	L 8·0
24	14·3	146·3	37·2	R *16·1*	R 7·0	R 8·0
				L *16·5*	L 7·0	L 7·5
25	14·5	152·6	54	R 18·2	R 7·5	R 8·5
				L 18·5	L 7·5	L 8·1
26	14·6	155	35·4	R 17·2	R 6·9	R 7·4
				L 17·5	L 7·0	L 7·5
27	14·6	147·5	41·7	R *16·0*	R 7·3	R 7·9
				L *16·0*	L 7·3	L 7·6
28	14·6	138·7	33·8	R *16·5*	R 7·0	R 7·4
				L 16·7	L 6·6	L 7·0
29	14·6	146·3	44·35	R 17·3	R 8·0	R 8·5
				L 17·3	L 7·6	L 8·4
30	14·6	148·8	40·4	R 17·6	R 7·8	R 8·2
				L 17·9	L 7·5	L 7·8
31	14·6	146·3	38·6	R *16·7*	R 7·0	R 7·3
				L *17·2*	L 7·0	L 7·2
32	14·6	153·8	39·5	R 17·3	R 7·0	R 7·2
				L 17·5	L 6·9	L 7·0
33	14·6	157·6	42·6	R 18·5	R 7·2	R 7·6
				L 18·7	L 7·3	L 7·7
34	14·8	150	44·9	R 17·8	R 7·8	R 8·3
				L 17·6	L 8·0	L 8·3
35	14·8	147·5	44·3	R 18·0	R 7·5	R 8·6
				L 18·5	L 7·5	L 7·8
36	14·9	142·5	33·6	R 16·9	R 7·2	R 7·6
				L 17·0	L 7·0	L 7·2
37	14·10	142·5	41·7	R 17·9	R 7·5	R 8·0
				L 18·3	L 7·5	L 7·8
38	15	174	61·05	R 19·5	R 8·4	R 9·4
				L 19·6	L 8·2	L 9·0
39	15	141·2	36	R *15·6*	R 6·2	R 7·0
				L *16·0*	L 6·5	L 7·2
40	15	158·9	46	R 18·5	R 7·5	R 8·0
				L 18·5	L 7·5	L 8·0
41	15	145	39·7	R 17·1	R 7·1	R 7·6
				L 17·3	L 7·1	L 7·5
42	15	166·6	58·2	R 20·0	R 8·2	R 9·9
				L 19·8	L 8·5	L 9·0
43	15	146·3	36·3	R 16·4	R 7·0	R 7·2
				L *16·5*	L 6·9	L 7·0
44	15	153·8	45·8	R *19·3*	R 8·0	R 8·5
				L 20·0	L 8·3	L 8·5

CASES	AGE	HEIGHT (CM.)	WEIGHT (KG.)	LENGTH OF HAND (CM.)	WIDTH OF HAND W1 (CM.)	WIDTH OF HAND W2 (CM.)
45	15	178	68·4	R 20·7	R 8·6	R 9·6
				L 21·0	L 8·5	L 9·2
46	15	157·6	52·6	R 19·0	R 7·6	R 9·0
				L 19·1	L 8·0	L 8·9
47	15	153	49	R 17·6	R 7·5	R 8·5
				L 18·1	L 7·8	L 8·3
48	15	162	76·6	R 18·3	R 8·0	R 9·0
				L 18·5	L 8·0	L 8·6
49	15	136	35·4	R 16·0	R 7·6	R 8·0
				L 16·2	L 7·1	L 7·6
50	15	166·5	51·7	R 19·8	R 8·6	R 9·3
				L 20·0	L 8·5	L 9·2
51	15	138	28·6	R 15·6	R 5·6	R 6·2
				L *16·0*	L 6·0	L 6·7
52	15	158	47·8	R 19·2	R 8·0	R 8·0
				L 19·1	L 8·0	L 7·8
53	15	151	47·7	R 18·0	R 7·2	R 7·8
				L 18·4	L 7·1	L 7·5
54	15·3	152	40·8	R 17·7	R 7·7	R 8·4
				L 18·0	L 7·7	L 7·4
55	15·3	151·4	45·1	R 18·5	R 8·0	R 9·0
				L 18·7	L 8·0	L 8·9
56	15·4	156	45·1	R 18·2	R 7·4	R 8·2
				L 18·3	L 7·5	L 8·0
57	15·6	143	36	R 17·2	R 7·0	R 7·7
				L 17·3	L 7·0	L 7·5
58	15·7	153	44	R 17·8	R 7·3	R 7·7
				L 17·8	L 7·5	L 7·6
59	16	158	46	R 19·5	R 7·5	R 8·3
				L 19·6	L 8·5	L 8·6
60	16	153·8	44	R 19·3	R 8·2	R 9·0
				L 19·5	L 8·0	L 8·9

HANDS OF NORMAL GIRLS

CASES	AGE	HEIGHT (CM.)	WEIGHT (KG.)	LENGTH OF HAND (CM.)	WIDTH OF HAND W1 (CM.)	WIDTH OF HAND W2 (CM.)
1	Almost 14	163	53·5	R 18·2	R 7·5	R 8·0
				L 17·9	L 7·4	L 7·9
2	,,	155	47·6	R 18·5	R 7·1	R 7·5
				L 18·5	L 7·1	L 7·4
3	,,	173	56·2	R 19·3	R 7·5	R 7·4
				L 20·0	L 7·3	L 7·7
4	,,	157	48·5	R 18·5	R 7·0	R 7·3
				L 18·4	L 7·0	L 7·1
5	,,	160	47·4	R 18·1	R 7·1	R 8·0
				L 18·4	L 7·1	L 7·6
6	,,	163	49·2	R 19·0	R 7·5	R 7·9
				L 19·4	L 7·5	L 7·5
7	,,	160	62·5	R 17·5	R 7·3	R 7·8
				L 17·9	L 7·1	L 7·6
8	,,	173	56·3	R 19·6	R 7·3	R 7·7
				L 19·4	L 7·3	L 7·7
9	,,	164	62·1	R 18·8	R 7·5	R 8·0
				L 19·2	L 7·4	L 7·8
10	,,	157	53·3	R 17·3	R 7·5	R 8·0
				L 17·6	L 7·5	L 7·8
11	,,	160	48·8	R 18·6	R 7·1	R 7·5
				L 19·1	L 7·1	L 7·3
12	,,	168	53·3	R 18·5	R 7·4	R 8·0
				L 18·8	L 7·2	L 7·5
13	,,	159	46·1	R 17·7	R 7·0	R 7·6
				L 17·6	L 7·0	L 7·5
14	,,	155	40·3	R 17·8	R 7·0	R 7·5
				L 18·1	L 7·3	L 7·5
15	14	156	49·4	R 18·0	R 7·5	R 8·1
				L 18·5	L 7·4	L 7·8
16	14	169	57·1	R 18·8	R 8·0	R 8·3
				L 19·2	L 7·5	L 8·0
17	14	169	62·0	R 19·4	R 7·6	R 8·5
				L 19·6	L 7·5	L 8·4
18	14	165	47·3	R 19·2	R 7·5	R 7·8
				L 19·6	L 7·4	L 7·6
19	14	161	57·1	R 18·0	R 7·5	R 8·0
				L 18·5	L 7·5	L 7·8
20	14	159	37·4	R 18·2	R 7·2	R 7·6
				L 18·5	L 7·2	L 7·6
21	14	150	43·0	R 17·6	R 7·7	R 8·2
				L 17·7	L 7·5	L 8·6

CASES	AGE	HEIGHT (CM.)	WEIGHT (KG.)	LENGTH OF HAND (CM.)	WIDTH OF HAND W1 (CM.)	WIDTH OF HAND W2 (CM.)
22	14	157	49·8	R 18·4	R 7·6	R 7·8
				L 18·0	L 7·6	L 7·8
23	14	169	52·3	R 18·2	R 7·5	R 7·9
				L 18·6	L 7·5	L 7·7
24	14	155	38·0	R 17·3	R 6·9	R 7·2
				L 17·4	L 6·6	L 7·1
25	14	164	51·7	R 18·5	R 7·5	R 7·7
				L 19·6	L 7·4	L 7·6
26	14	160	59·8	R 17·0	R 7·0	R 7·7
				L 17·4	L 7·0	L 7·5
27	14	167	52·0	R 19.3	R 7·5	R 8·0
				L 19·5	L 7·3	L 7·6
28	14	160	46·7	R 17·7	R 7·5	R 7·9
				L 18·2	L 7·4	L 7·5
29	14	160	59·8	R 17·8	R 7·5	R 8·2
				L 18·1	L 7·4	L 7·6
30	14	160	53·0	R 17·2	R 8·2	R 8·6
				L 17·4	L 7·5	L 7·9
31	14	169	65·3	R 19·2	R 7·5	R 8·2
				L 18·5	L 7·3	L 7·8
32	14	164	54·2	R 18·8	R 7·4	R 7·9
				L 18·8	L 7·4	L 7·6
33	14	160	60·2	R 18·9	R 8·3	R 9·0
				L 18·5	L 7·6	L 8·4
34	13	154	48·7	R 17·5	R 7·0	R 7·6
				L 17·5	L 7·0	L 7·5
35	14	159	48·5	R 18·3	R 7·6	R 8·2
				L 18·5	L 7·5	L 7·9
36	14	159	44·9	R 18·0	R 7·2	R 7·5
				L 18·0	L 7·2	L 7·3
37	14	153	57·2	R 17·5	R 7·2	R 7·8
				L 17·8	L 7·1	L 7·5
38	14	159	47·3	R 18·0	R 7·4	R 7·7
				L 18·5	L 7·5	L 7·6
39	14	164	50·8	R 19·3	R 7·5	R 7·8
				L 19·0	L 7·4	L 7·7
40	14	168	50·5	R 19·6	R 7·6	R 8·1
				L 20·0	L 7·5	L 7·8
41	14	159	60·4	R 17·8	R 7·5	R 8·0
				L 18·0	L 7·5	L 8·0
42	14	161	72·0	R 18·1	R 7·5	R 7·9
				L 18·3	L 7·2	L 7·6
43	14	163	49·3	R 18·3	R 7·6	R 7·8
				L 18·5	L 7·3	L 7·5
44	14	158	52·3	R 17·5	R 7·0	R 7·6
				L 17·5	L 7·0	L 7·5

CASES	AGE	HEIGHT (CM.)	WEIGHT (KG.)	LENGTH OF HAND (CM.)	WIDTH OF HAND W1 (CM.)	WIDTH OF HAND W2 (CM.)
45	14	162	59·4	R 18·8	R 7·4	R 7·9
				L 18·8	L 7·3	L 7·6
46	14	159	49·4	R 17·8	R 7·7	R 7·8
				L 17·9	L 7·5	L 7·8
47	14	163	59·6	R 18·1	R 7·5	R 8·1
				L 17·9	L 7·4	L 7·9
48	14	163	52·3	R 18·5	R 7·0	R 7·3
				L 18·4	L 7·0	L 7·1
49	14	167	69·0	R 17·8	R 7·0	R 7·5
				L 18·0	L 6·9	L 7·3
50	14	167	66·1	R 18·9	R 8·3	R 9·0
				L 18·5	L 7·6	L 8·4
51	14	161	57·2	R 18·5	R 7·5	R 7·9
				L 19·0	L 7·4	L 7·8
52	14	174	73·0	R 18·8	R 7·3	R 7·6
				L 18·8	L 7·1	L 7·4
53	14	167	60·8	R 18·6	R 7·2	R 7·9
				L 18·1	L 7·3	L 7·7
54	15	158	61·2	R 17·8	R 7·5	R 7·9
				L 17·9	L 7·3	L 7·6
55	15	158	52·4	R 18·2	R 7·8	R 8·6
				L 18·5	L 7·5	L 8·5
56	15	155	46·4	R 17·4	R 7·0	R 7·8
				L 17·7	L 7·0	L 7·8
57	15	158	61·2	R 17·8	R 7·5	R 7·9
				L 17·9	L 7·3	L 7·6
58	15	167	60·8	R 18·0	R 7·2	R 7·9
				L 18·0	L 7·3	L 7·7
59	15	175	56·5	R 18·2	R 7·4	R 7·6
				L 18·5	L 7·0	L 7·3
60	15	171	63·5	R 19·6	R 7·5	R 8·1
				L 19·6	L 7·2	L 7·8

HANDS OF MENTALLY DEFECTIVE GIRLS

CASES	AGE	HEIGHT (CM.)	WEIGHT (KG.)	LENGTH OF HAND (CM.)	WIDTH OF HAND W1 (CM.)	WIDTH OF HAND W2 (CM.)
1	13	152	44·4	R 17·4	R 7·9	R 7·5
				L 17·5	L 7·0	L 7·7
2	14	158	51·7	R 18·6	R 7·6	R 8·3
				L 19·0	L 7·3	L 8·2
3	14	148	41·7	R *16·2*	R 6·5	R 7·0
				L *16·9*	L 6·4	L 7·0
4	14	146	40·7	R 17·6	R 7·0	R 7·3
				L 17·6	L 7·0	L 7·2
5	14	162	51·0	R 19·6	R 7·9	R 8·3
				L 19·8	L 7·4	L 8·0
6	14	150	46·9	R 18·8	R 7·2	R 7·7
				L 18·8	L 7·2	L 7·6
7	14	151	44·0	R 17·6	R 7·5	R 8·0
				L 17·8	L 7·3	L 7·9
8	14	160	54·9	R 19·0	R 7·9	R 8·1
				L 19·2	L 7·9	L 8·0
9	14	159	52·0	R 18·8	R 7·2	R 7·8
				L 18·8	L 7·2	L 7·8
10	14	155	44·0	R 18·5	R 7·6	R 7·1
				L 18·5	L 8·0	L 7·2
11	14	153	41·2	R 18·5	R 7·9	R 7·3
				L 19·0	L 7·3	L 7·0
12	14	141	46·7	R *16·2*	R 7·2	R 7·4
				L *16·5*	L 6·9	L 6·9
13	14	160	45·8	R 17·7	R 6·7	R 7·2
				L 17·5	L 6·3	L 7·1
14	14	151	41·0	R 17·5	R 7·0	R 7·2
				L 17·7	L 6·9	L 7·2
15	14	147	40·6	R *15·8*	R 6·6	R 7·0
				L *16·1*	L 6·6	L 6·9
16	14	141	36·3	R 17·4	R 6·8	R 7·0
				L 17·5	L 6·9	L 7·4
17	14	153	52·2	R 18·2	R 7·5	R 8·0
				L 18·3	L 7·3	L 8·2
18	14	146	41·3	R 17·6	R 7·2	R 7·3
				L 17·8	L 7·2	L 7·0
19	15	155	54·6	R 19·0	R 7·7	R 8·5
				L 18·7	L 7·5	L 8·1
20	15	143	44·4	R *16·6*	R 7·4	R 7·9
				L *16·7*	L 7·2	L 7·6
21	15	144	41·7	R 18·2	R 7·0	R 7·2
				L 18·4	L 6·9	L 7·0

CASES	AGE	HEIGHT (CM.)	WEIGHT (KG.)	LENGTH OF HAND (CM.)	WIDTH OF HAND W1 (CM.)	WIDTH OF HAND W2 (CM.)
22	15	142	48·5	R *16·0*	R 7·1	R 7·4
				L *16·4*	L 7·0	L 7·2
23	15	157	49·9	R 18·6	R 7·1	R 7·6
				L 18·7	L 7·1	L 7·3
24	15	160	48·3	R 19·1	R 7·2	R 7·8
				L 19·5	L 7·2	L 7·6
25	15	149	49·0	R 17·5	R 7·2	R 7·9
				L 17·9	L 7·1	L 7·8
26	15	153	49·0	R 17·1	R 7·1	R 7·8
				L 17·2	L 7·0	L 7·5
27	15	153	49·0	R 17·1	R 7·3	R 7·8
				L 17·4	L 7·3	L 7·5
28	15	160	48·7	R 17·4	R 6·9	R 7·0
				L 17·8	L 6·8	L 7·0
29	15	158	43·5	R 18·5	R 7·2	R 7·6
				L 18·5	L 7·5	L 7·8
30	15	157	47·6	R 17·8	R 6·8	R 7·2
				L 18·2	L 6·7	L 7·1
31	15	157	50·4	R 18·8	R 7·5	R 7·8
				L 19·2	L 7·1	L 7·6
32	15	161	51·7	R 19·8	R 7·4	R 8·3
				L 19·6	L 7·1	L 8·5
33	15	143	39·9	R 17·2	R 6·9	R 7·4
				L 17·3	L 6·9	L 7·4
34	15	160	49·4	R 18·6	R 7·5	R 8·0
				L 18·7	L 7·5	L 8·0
35	15	162	50·8	R 16·6	R 7·6	R 7·0
				L 17·1	L 7·5	L 7·0
36	15	155	53·4	R 17·8	R 8·0	R 7·4
				L 17·8	L 8·2	L 7·2
37	15	160	49·0	R 17·6	R 7·0	R 7·4
				L 17·9	L 6·5	L 7·1
38	15	152	37·0	R 17·5	R 6·9	R 7·1
				L 17·5	L 6·9	L 6·9
39	15	153	54·0	R 18·5	R 7·6	R 8·4
				L 18·8	L 7·4	L 8·0
40	15	146	42·6	R 17·0	R 6·9	R 7·3
				L 17·0	L 6·9	L 7·2
41	15	140	41·1	R 17·3	R 7·8	R 8·4
				L 17·6	L 8·0	L 8·5
42	15	164	47·6	R 19·1	R 8·1	R 8·5
				L 19·4	L 8·1	L 8·2
43	15	157	51·3	R 18·9	R 7·8	R 8·1
				L 19·1	L 7·4	L 8·1
44	15	148	39·0	R *16·6*	R 6·9	R 7·0
				L *16·9*	L 6·5	L 6·6

CASES	AGE	HEIGHT (CM.)	WEIGHT (KG.)	LENGTH OF HAND (CM.)	WIDTH OF HAND W1 (CM.)	WIDTH OF HAND W2 (CM.)
45	15	156	47·2	R 18·4	R 7·0	R 7·7
				L 18·7	L 6·9	L 7·5
46	15	167	56·7	R 19·5	R 7·5	R 8·0
				L 19·6	L 7·3	L 8·2
47	15	148	49·2	R 17·6	R 7·0	R 7·6
				L 17·3	L 7·0	L 7·5
48	15	142	39·6	R 16·6	R 6·9	R 7·3
				L 16·8	L 6·9	L 7·1
49	15	148	49·0	R 17·2	R 7·1	R 7·5
				L 17·4	L 7·1	L 7·3
50	15	148	42·0	R 17·6	R 7·3	R 7·1
				L 17·6	L 7·2	L 7·0
51	15	157	49·9	R 18·6	R 7·6	R 7·1
				L 18·7	L 7·3	L 7·1
52	15	155	55·1	R 17·2	R 7·4	R 8·0
				L 17·2	L 7·3	L 8·1
53	15	147	46·2	R 16·7	R 7·1	R 7·4
				L 16·9	L 7·0	L 7·3
54	16	152	54·7	R 17·1	R 7·3	R 7·8
				L 17·5	L 7·2	L 7·6
55	16	167	66·2	R 19·8	R 8·1	R 8·9
				L 20·0	L 8·0	L 8·6
56	16	165	63·4	R 19·8	R 7·2	R 8·0
				L 20·0	L 7·1	L 7·9
57	16	147	45·1	R *16·4*	R 7·1	R 7·6
				L *16·7*	L 6·6	L 7·4
58	16	153	54·0	R 18·6	R 7·2	R 8·0
				L 18·9	L 7·2	L 8·0
59	16	142	34·9	R 16·8	R 7·0	R 7·5
				L 17·4	L 6·9	L 7·1
60	16	150	54·0	R 17·1	R 7·1	R 7·8
				L 17·5	L 7·0	L 7·5

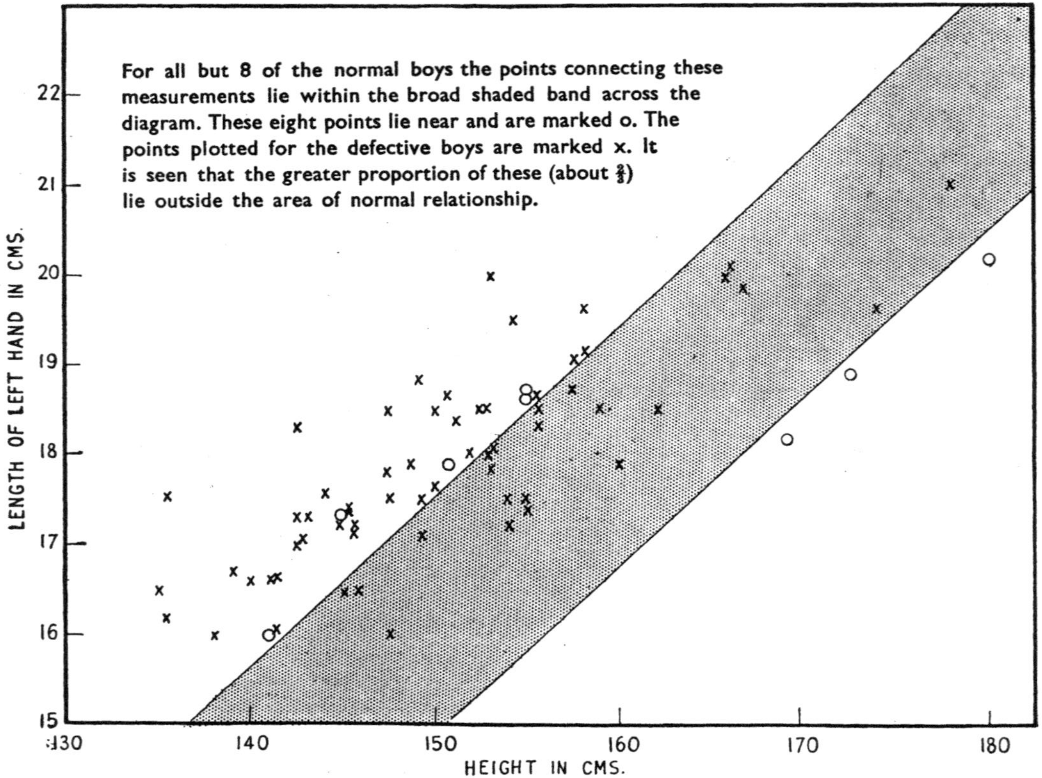

GRAPH I SHOWING HOW THE HEIGHT OF THE BODY CORRELATES WITH THE LENGTH OF THE LEFT HAND FOR NORMAL AND FOR MENTALLY DEFECTIVE BOYS (*See Tables*)

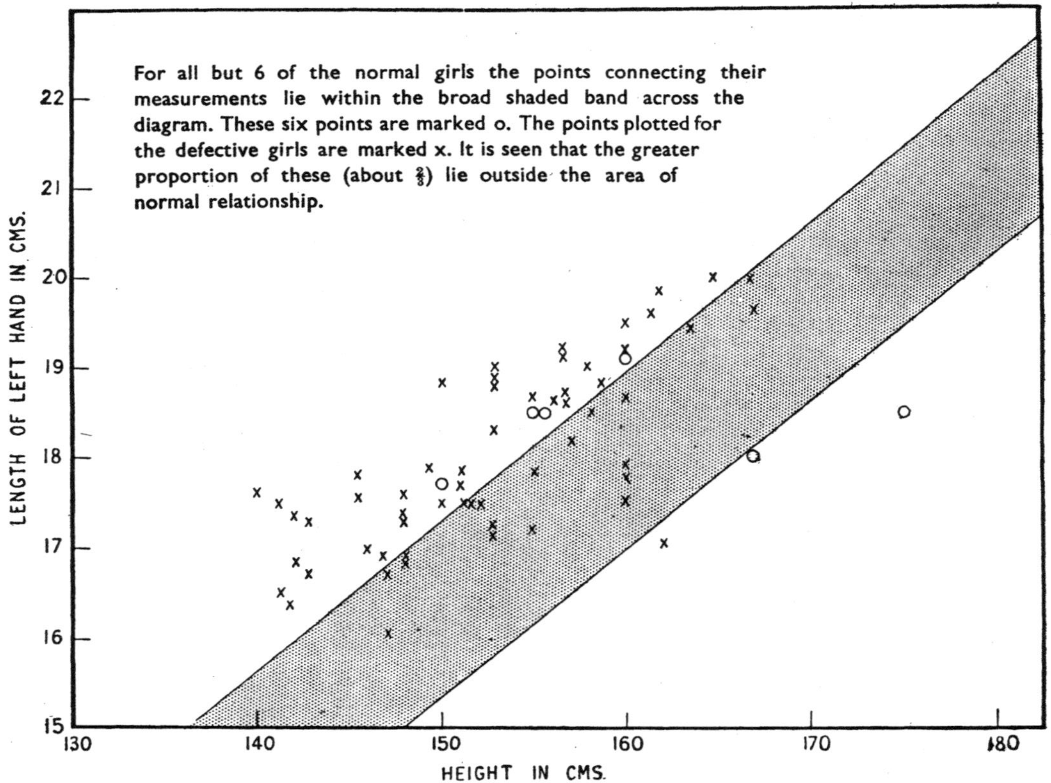

GRAPH 2 SHOWING HOW THE HEIGHT OF THE BODY CORRELATES WITH THE LENGTH OF THE LEFT HAND FOR NORMAL AND FOR MENTALLY DEFECTIVE GIRLS (*See Tables*)

THE MEASUREMENTS

I have noted in every case under investigation the following measurements:

1. HEIGHT AND WEIGHT

and

2. LENGTH OF BOTH HANDS
3. WIDTH OF BOTH HANDS.

The very simple lists of these measurements show at a glance the following results:

A. There is a retardment of growth in the high-grade defective child; at the same time one finds in a significant number the weight below normal. The hands are acromicric in 7 cases of both mentally defective boys and girls.

This result is neither new nor striking. It is what one would expect. Mental retardment goes with physical retardment, as a rule. There are a few cases where the opposite occurs: namely a precocious physical development and the early attainment of a tall figure and excellent muscular build. But these cases are the rare exceptions in both high-grade boys and girls.

B. In a great number of cases of mental deficiency both the length and the width of the hands are not correlated with the height and weight of the children concerned. One finds abnormally small hands in tall children and narrow and thin ones in children of normal weight. This is an unexpected result, in any case for the commonsense idea of bodily proportions. There are, of course, a number of cases where the length and the width are in proportion to the whole physique, but the opposite condition was conspicuous. Very long hands are, however, absent in very small children, but there are some comparatively very broad hands in children of a comparatively very low weight.

These findings are of interest and importance. The fact that the volume of the body is not correlated with the volume of the hands can be easily understood through

1. Sheldon's explanation of the voluminous endomorph person, whose sphericity is due mainly to a large amount of fat tissue and the tendency to water retention in the body. The muscular development of these people is rather poor, which explains the weakness of their palmar muscles.

2. The described 'inconsequence' of nature can be even better understood, when one considers the functioning of the endocrine

glands in their relation to growth. In both hypothyroidism and hypopituitarism one finds an early closure of the epiphyses of the bones of the fingers. While hypothyroic people are small in height, when the condition is congenital, hypopituitary individuals are often of medium height or even comparatively tall. I am speaking of that type of hypopituitarism which goes with obesity, and which is called in its extreme manifestation the Froehlich syndrome. The hands which go with this condition are most often small, often slender and elegant, quite in contrast to the whole physique. The feet follow the same design as the hands and it is well known that these 'heavy' people are often excellent dancers and surprisingly light and graceful in their movements. Hypopituitarism of this type is however unusual in high-grade defectives as I have already mentioned earlier in this book. We find, of course, similar conditions of disproportion between measurements of the hand and height and weight in normal children, though not in the same degree.* One must assume that in all cases where there occurs an obvious disproportion between hand measurements and other body measurements there will be a definite endocrine factor responsible for the condition described. It is also worth mentioning that fat deposits are found to a much slighter degree on the palm than on the fingers, the basal phalanxes in particular. The sausage-shaped fingers of hypopituitary and hypothyroic people are well known to the physician, particularly the endocrinologist.

Long and thin hands are the rule in the overgrown slim adolescent with a tendency to subgonadism, in which cases the measurements of the hand are in line with the general physical build.

In 15 of the high-grade boys investigated I observed long hands of the elementary regressive type together with a comparatively small physique and often coupled with secondary feminine sex characteristics. These hands are also out of proportion in relation to the whole physique and their shape is due to hypopituitary function of a kind different from that described in the preceding paragraph. This kind of pituitary infantilism, called the Levy Lorraine syndrome, seems to occur far more often in boys than in girls, in any case as far as my experience goes.

This result is of diagnostic value: it means that the hand shows signs of endocrine dysfunction of a congenital kind at an early stage,

* The two graphs on pp. 174 and 175 show the far greater disproportion between the length of the hand and the height of the body in mentally defective than in normal children.

when there may be no other definite physical symptoms visible and perhaps no clinical symptoms at all. The importance of detecting such endocrine condition in early childhood before the manifestation of any physical or mental disability, is of scientific as well as practical importance.

THE TABLES

I have tried to arrange my material according to the most significant data in the simplest way possible and have listed it in three tables. These refer to certain hand-traits, which I have investigated in both groups of children, the normal and the mental defective. The tables correspond to those of the previous chapters and show the numerical incidence of the six principal hand-types in the two groups of children (Table I), of characteristics of the fingers and nails and of the crease-lines and papillary ridges (Tables II and III).

TABLE I

shows the Numerical Incidence of Certain Hand-Types in 240 Children, 120 Normal and 120 Mental Defective

	ELEMENTARY REGRESSIVE LONG FINGERS	ELEMENTARY REGRESSIVE SHORT FINGERS	ELEMENTARY SIMPLE	FLESHY-MOTOR	BONY-MOTOR	SENSITIVE SMALL	SENSITIVE LONG	ACROMICRY
60 NORMAL BOYS	4	1	10	10	18	9	7	1
60 MENTALLY DEFECTIVE BOYS	20	15	6	3	4	4	1	7
60 NORMAL GIRLS	0	3	5	17	10	11	14	0
60 MENTALLY DEFECTIVE GIRLS	6	6	2	2	8	18	11	7

I have omitted in this study to make a statistical survey of the physical qualities as I have dealt with this aspect in the chapter 'The Hand of the Mental Defective'. Though the physical qualities of the hand have a certain significance for the emotional make-up and particularly for the inhibitive factor, their assessment is, apart from the degree of flexibility, rather difficult in 'normal' children, who are more conscious of the test situation and have therefore often moist and sometimes pale hands because of excitement and anxiety.

TABLE II

shows the Numerical Incidence of Certain Characteristics of Fingers and Nails in 240 Children, 120 Normal and 120 Mental Defective

	FINGER-TIPS				NAILS				THUMB			FIFTH FINGER			HUMAN DIGITAL FORMULA	ANKYLOSIS
	SPATULATE	TAPERED	SQUARE	OVERLONG	LONG	MEDIUM	SMALL	RUDI-MENTARY	TOO LONG	TOO SHORT	MISSHAPEN	TOO LONG	TOO SHORT	MISSHAPEN		
60 NORMAL BOYS	12	15	33	14	19	30	11	0	1	9	8	2	2	7	5	0
60 MENTALLY-DEFECTIVE BOYS	30	8	22	26	4	21	20	15	4	12	17	4	14	14	0	1
60 NORMAL GIRLS	14	25	21	8	23	27	10	0	1	5	3	13	4	3	8	0
60 MENTALLY-DEFECTIVE GIRLS	22	18	20	4	3	14	27	16	1	19	25	0	17	20	0	0

TABLE III

shows the Numerical Incidence of Abnormalities of the Crease-Lines and the Papillary Ridges in 120 Normal and 120 Mentally Defective Children

	ABNORMALITIES OF CREASE-LINES				*ABNORMALITIES OF THE PAPILLARY RIDGES*				
	SIMIAN LINE	LONG LONGI-TUDINAL LINE MISSING	FAULTS UPPER TRANSVERSE CREASE	FAULTS LOWER TRANSVERSE CREASE	WHORLS OR LOOPS THENAR EMINENCE	WHORLS OR LOOPS HYPO-THENAR EMINENCE	ARC-TRIANGLE PATTERN ON ONE FINGER	ARC-TRIANGLE PATTERN ON SEVERAL FINGERS	ARC-TRIANGLE PATTERN ON ALL FINGERS
60 NORMAL BOYS	3	6	13	23	7	21	7	7	0
60 MENTALLY DEFECTIVE BOYS	8	14	32	36	12	23	12	16	2
60 NORMAL GIRLS	3	7	17	20	0	6	10	13	1
60 MENTALLY DEFECTIVE GIRLS	5	7	18	38	6	17	10	18	5

It is important to mention that the children whose hands are investigated come from similar strata of society: they belong mostly either to the working class or lower middle class. It seemed desirable to draw my material for this study from a homogeneous milieu, where nutritional and living conditions are on a fairly equal level. The nutritional factor especially has a certain relevance to physical constitution. It must also be pointed out that a number of the children with mental deficiency came from families where the parents were divorced or separated or where either the father or the mother were 'unsatisfactory' for various reasons.

TABLE I

This table shows the numerical incidence of the six main hand-types in the two groups of children.

I have made a subdivision of the elementary regressive hands into those with long and those with short fingers, as the former hand-type was frequently observed in high-grade boys.

The elementary regressive hand was the most frequent one in high-grade defective boys; it was found in 35 out of 60 compared with 5 cases of normally intelligent boys. The variety with long fingers was predominant with 20 cases, while the elementary hand with short fingers occurred 15 times in this group. Among the normal boys the latter variety was observed in one case only. One of the most interesting results of this table is the fact that high-grade girls possessed elementary regressive hands in a far lesser number, namely in only 12 cases altogether, 6 to each of the two varieties.

The corresponding group of normally intelligent girls had elementary regressive hands in 3 cases only, all of which were hands with short fingers.

The elementary simple hand occurred most often in normal boys, in 10 cases compared with only 6 in mentally defective boys. This hand-type was 5 times present among normal girls and only twice in the corresponding group of high-grade girls.

The fleshy motor hand, which is much akin to the elementary simple hand, occurred most often in normal girls, namely in 17 cases, compared with only 2 in the group of mentally defective girls. The same hand-type was found in 10 of the normal and 6 of the high-grade boys.

The incidence of the bony motor type of hand among normal and among 'defective' children showed little difference: among the normal boys investigated there were 18 cases, compared with 4 in the 'high-grade' boys; while with the girls 10 of the normal children had this type of hand compared with 8 of the mental defectives.

Another important result to be seen from this table is that the sensitive small type of hand was most frequently found in high-grade girls, and least often in high-grade boys. It occurred 11 times in normally intelligent girls and 9 times in normally intelligent boys.

The sensitive long hand was most often present in the normal girls investigated, where it occurred 14 times, compared with 11 times in mentally defective girls. It was present only once in the group of the high-grade boys compared with 7 in that of the normal boys.

Acromicric hands, which represent an exaggerated form of the sensitive small hand, were found in 7 cases of the mentally defective boys and girls, and once among the normal boys.

DISCUSSION OF THE RESULTS

I. HANDS OF NORMALLY INTELLIGENT CHILDREN

Two things are striking about the hands of the normal children investigated: the very frequent occurrence of the motor types of hand and a different distribution of hand-types in regard to sex.

The fleshy and bony motor hand as well as the elementary simple one can be called 'active' hands in order to differentiate them from the more receptive or 'passive' hands, such as the sensitive small and sensitive long one. 38 of the boys had 'active' hands compared with 16 who had 'sensitive' hands. The bony motor hand takes first place, the elementary simple and fleshy motor ones in equal numbers come second and third, and elementary regressive hands are the exception. The former hand-types can be considered as the most-developed and mature of the six hand-types because of their structural and muscular development. It is interesting to note that the bony motor hands were more often found than the fleshy motor ones, which may be a particular 'racial' characteristic of the English. I think that the comparatively high frequency of the elementary simple hand has a social significance, as these boys come from the working class. It is among workmen and peasants that one finds this hand most often.

The 'active' hands were also prevalent among the girls, but the distribution differed insofar as the fleshy motor hand outnumbered the bony motor one by 7 and the elementary simple hand by 12. This corresponds to a general difference in constitutional build of the sexes; the greater angularity of the male and the greater roundness of the female. Though the predominance of the 'active' hands brings normal boys and girls in line with each other in contrast to mental defectives, there is a marked difference in the number of 'sensitive' hands in the sexes. This is, of course, to be expected: the sensitive long hand was 14 times found in girls and only 7 times in boys and the sensitive small hand in 11 girls compared with 9 boys. I have said in my earlier writings that the sensitive hand is a woman's hand, and particularly the small variety. This statement is only partly justified according to this table. It is certainly true that the passive hand is of a feminine character and occurs less often in men, but the sensitive small hand was found fairly often in these boys. The number of cases is however too small to come to any conclusion on this point. There was one case of acromicry among the normal boys.

2. HANDS OF HIGH-GRADE DEFECTIVE CHILDREN

One glance at the numbers relating to the distribution of hand-types in mental defective children shows a striking difference from those of the normal group. But one similarity exists between the two groups, and that is the marked difference of hand-types in the sexes.

THE ELEMENTARY REGRESSIVE HAND

may be called the hand most characteristic of the high-grade boy and one would expect to find the same in feebleminded girls. But that is not so.

DISTRIBUTION OF 'NORMAL' AND ABNORMAL HANDS

There are only 13 'active' hands in the group of boys; 4 bony motor, 6 elementary simple and 3 fleshy motor ones, compared with 12 'passive' or sensitive hands. The majority of hands was 'abnormal' and the most frequent abnormality in high-grade defective boys was the elementary regressive hand.

THIS FINDING IS IN ITSELF DIRECT EVIDENCE FOR A CONSTITUTIONAL FACTOR RELATED TO CERTAIN DEFINITE AND ABNORMAL CHARACTERISTICS IN THE DEFECTIVE PERSONALITY. I am using the word defective instead of mentally defective as the constitutional factor permeates the whole of the psycho-physical organism and does not only affect the intellectual capacity.

This result seems to me of great scientific and diagnostic value.

If certain hand-types together with characteristic hand-traits, which are mostly congenital, reveal the tendency to some form or other of abnormality either on the mental or on the emotional plane or both, the child may get the right treatment at an early moment. I remind the reader here that the elementary regressive hand is not only characteristic of the high-grade mental defective, but is also frequently found in intelligent but emotionally unbalanced people. This restriction of diagnosis must always be kept in mind. It also need hardly be mentioned that any diagnosis from hand-traits alone would be a grave mistake. The hand can serve as a most useful guide to diagnosis but, as I have repeatedly said, its indications must be confirmed by other tests.

This reserve does not take anything away from the diagnostic value of the hand; it only puts the latter into the right place in the field of diagnosis.

The fact that the majority of these boys possessed elementary regressive hands with long fingers corresponds to the finding of my previous investigation on the same subject, on which I have reported in the chapter 'The Hand of the Mental Defective'. This type of hand with long fingers, a well-marked thenar and hypothenar eminence and a rather thin and flat distal part of the palm is characteristic of the high-grade defective with a schizothyme make-up.

There were 7 cases of acromicry among this group of boys, an abnormality that generally occurs in more or less dwarfed individuals.

Among the mentally defective girls the elementary regressive hand occurred in only 12 cases and equal numbers were found of each variety of this hand-type. This is a puzzling result which I shall presently try to explain. Though this is a comparatively small number as compared with what was found among high-grade boys, it is still significantly larger than for normal girls.

Abnormal hand-types were far less frequent among high-grade girls than among boys, even if we take into consideration the 7 acromicric pairs of hands which would bring the total to 19, compared with 42 among the boys. One glance at the numbers in this table is enough to show that the majority of the high-grade defective girls possessed 'passive' hands. There were 36 pairs of the 'sensitive' groups of hand, in which the 7 cases of acromicry were included.

The proportion of 'active' hands was negligible in these girls. The fleshy motor hand occurred only twice and so did the elementary simple one. The 'active' types of hand occurred almost three times as often in the normal girls included in this investigation.

Among the 'passive' hands the sensitive small one was predominant. *It seems likely that the small sensitive type of hand has a special significance as related to the type of girl one finds in an approved school or in a Colony for mental defectives.*

How can these results, if at all, be explained?

Anyone who has had to deal with high-grade defective girls will know that they have, as a rule, a highly excitable nervous system, an 'emotional' temperament, and that they are extremely fond of the boys, for which reason they generally get into trouble. Nymphomania is not rare among these girls. There is an hysterical streak in most of them. Certainly, the dispositions of temperament and personality which bring high-grade girls into conflict with society are of a different kind from those which are responsible for the failings of the boys. It is therefore logical to assume that we find a particular make-up, different from that of feeble-minded boys, in high-grade difficult girls. In them, it is mostly an offence of a sexual nature that calls for special education and constant supervision. The sensitive small hand was also common in the 'normal' group of girls, though the sensitive long hand occurred in them more often. The small sensitive hand goes with an emotionally unstable temperament and an easy disintegration of personality. One finds this type of hand frequently in women with paraphrenia and outside the Mental Hospital in women who suffer from neuroses of an obsessional reaction type. The small sensitive hand goes most often with emotional and physical infantilism. If one takes these characteristics into account, it seems obvious that the emotional disturbances of girls with this kind of hand is on a less severe level than those of their male counterparts. And this corresponds to the psychiatric diagnoses of these cases. The seven cases of acromicry refer to severe psychophysical infantilism. The kinds of temperament and personality which go with feeblemindedness and character disturbance in boys and in girls are not the same. *It is an interesting finding that this difference of temperament can be detected from hand-traits.*

It is equally interesting to note *that the nature of the possible disturbances that bring these girls into difficulties can be detected from hand-traits.*

On the whole, one finds a stronger tendency towards extraversion in the high-grade girl than in the high-grade boy, a finding that also corresponds to psychiatric observation.

Those girls who had elementary regressive hands were the more interesting cases because of the more complex personality disorders they were suffering from.

Though the difference between the hands of high-grade and normal girls is less dramatic than in the cases of normal and defective boys, it is evident that their hands show their mental and emotional disability. But as with other methods of diagnosis, more detailed findings have to be taken into account before any final indication of 'deficiency' of personality can be assessed.

TABLE II

I have listed in this table the most significant attributes of fingers and nails. I have differentiated the fingers according to the shape of their terminal phalanxes and also listed the number of overlong finger-tips. The size of the nails depends, as a rule, on the size of the fingers, which in turn depends on the length of the whole hand. Long, medium-sized and small nails are 'normal' nails, but the rudimentary nail, which means, as the term implies, an undeveloped, extremely small nail, is an abnormality.

The length and the shape of the marginal digits have been particularly observed and are mirrored in the data of this table.

While the human digital formula occurred comparatively often in schizophrenic patients, it is a rarity in mental defectives, which is borne out by this table.

Ankylosis is frequent in all types of psychosis, particularly in schizophrenia and mental deficiency, but occurred only once in the children of this investigation, in one mentally defective boy.

In order to show at this juncture how detailed analysis can help towards a differential diagnosis of mental and moral deficiency, I am altering the order in which I discuss results and am beginning with the hands of high-grade defective girls.

Let us look at the characteristics of the marginal fingers first: the thumb and the little finger. It is well known that these are the most liable to bear genetic stigmata. The 'clubbed' thumb as well as the misshapen little finger occur very often in the hands of mental defectives and their presence in the hands of mongolian imbeciles is recognized as one of the most characteristic marks in this condition. The abnormal shape of both fingers generally goes with their abnormal shortness. The resemblance of the minute, clubbed thumb to that of monkeys and apes is obvious, and one can therefore speak of a simian thumb as one speaks of a simian line, both of which appear to be hereditary and have therefore genetic significance.

Both the clubbed thumb and the misshapen little finger also occur sometimes in intelligent people and to make a diagnosis of mental deficiency on account of these hand-traits alone would be stupid to say the least, and only when there are other significant hand-features present is it safe to assume that the person concerned possesses certain constitutional tendencies which point to mental and emotional deficiency. I must emphasize once more that hand-diagnosis must be confirmed by other diagnostic estimates, but that the hand can lead to the discovery of psycho-pathological conditions at an age when the abnormality could not be traced otherwise.

The hands of the high-grade girls showed the greatest number of an abnormally shaped thumb, which was in 25 cases a clubbed and 'rigid' one. In 19 cases it was abnormally short.

The presence of such a thumb in the elementary regressive type of hand is less significant for diagnosis than in the sensitive small hand, where its occurrence is very rare. It was mostly in the latter type of hand that abnormal thumbs were found, a fact which increases their diagnostic value.

The Misshapen Little Finger, which also occurred most frequently in high-grade girls, is more 'enigmatic' for diagnostic purposes than the thumb. It is well known that the fifth finger is too short and often deformed in certain endocrine conditions, those which sometimes go with mental deficiency and those which do not. The diagnostic value of this item is therefore in itself minimal, but, in connexion with other 'degenerative' hand-traits, it gives some confirmation towards a specified diagnosis.

Another item, that seems to be characteristic of mental deficiency, namely *the Rudimentary Nail*, was found most often in this group, in 16 cases, compared with 15 in high-grade boys. This type of nail was never present in the normal boys and girls investigated.

Now then, if one discovers in a hand, the general shape of which already suggests abnormality, a clubbed and abnormally short thumb, a misshapen fifth finger and rudimentary nails, the suspicion of mental deficiency gets more and more justified.

In regard to the other items of this table only two are interesting for diagnostic reasons.

1. *The Overlong Finger-tip* and

2. The Spatulate Finger-tip

First of all, what is an overlong finger-tip? In order to establish an idea of the normal length of the finger-tips one must measure a number of hands and find out if there are definite proportions between the whole length of the fingers and the terminal phalanxes. I have measured the hands, fingers and finger-tips of all subjects under investigation in these clinical studies and the result for the point in question is this:

The longest finger is the medius finger, which at the same time can be called the most stable finger of the hand. If one takes the ratio of the length of the middle finger to the length of its terminal phalanx, one gets a measure, which is characteristic for the length of the finger-tips as a whole. In hands with finger-tips of normal length this ratio lies between 3·8 and 4·0. If it falls below this measure, the finger-tips are overlong.

I had noted a conspicuous length of the terminal phalanxes in the hands of high-grade defective boys in an earlier study and this observation gave me cause to believe that there is a correlation between this hand-trait and mental deficiency.

It is therefore most interesting to note that while the high-grade boys possessed overlong finger-tips in 26 cases, the feebleminded girls showed the same trait in only 4 cases and are last on the list in this respect of all the children.

Obviously there cannot be any correlation between overlong finger-tips and mental deficiency as such. As one sees the same hand-trait frequently in schizophrenics, it seems reasonable to assume that overlong finger-tips are significant rather of certain characteristics of temperament than of mentality. The most plausible suggestion is that the overlong finger-tip designates an overaccentuated inhibitive factor which produces abnormal introverted tendencies and develops a schizothyme personality. Indeed, this interpretation tallies with the personality of many of the high-grade boys I have met.

I have mentioned that the girls of the Colony were of a different temperament. An hysterical reaction type was predominant among them; and it is a well-known fact that hysteria commonly occurs among people with strong extravert tendencies. This may explain the absence of this hand-trait in the high-grade girls investigated.

Overlong finger-tips occurred in 14 normal boys and 8 normal girls, which is another proof that the correlation is with temperament rather than intelligence. The endocrine factor cannot be overlooked either. It is also common knowledge that hypogonadic hands have very long fingers with frequently overlong finger-tips. The closure of the epiphyses is delayed in these cases which explains the condition described.

The Spatulate Finger-tip is another item which occurs frequently in the hands of mental defectives. It is predominant in the high-grade boys; the girls take second place in this respect; but the spatulate finger-tip is the most frequent one in the hands of both sexes.

This is not so with normal children. The square finger-tip occurs most often in the boys, and the tapered one in the girls. The spatulate finger-tip comes last in both sexes.

I repeat here what I have mentioned earlier in this book: spatulate finger-tips are usual in the elementary regressive type of hand so characteristic of mental defectives, which explains the prevalence of this hand-trait in the high-grade children investigated.

The following hand-traits discussed are of less importance for this study but deserve mention.

The Length of the Nails

I have spoken of the *rudimentary nail* in the preceding discussion, as being a definite characteristic of the hand in mental deficiency. But there is also a marked general difference in the size of normal nails in normal and high-grade children.

Long, medium-sized and small nails occur in the hands of both normal and 'abnormal' people, but according to this table the prevalence of the long nail is found in the normal groups, while the prevalence of the short nail occurred in the high-grade groups.

The Human Digital Formula

The human digital formula is always rarer than the simian digital formula in both normal and pathological cases and this general statement is borne out by this particular study. Only 5 boys and 8 girls out of 120 normally intelligent children had a human digital formula which never occurred among feeble-minded children of either sex.

TABLE III

This table shows the numerical incidence of abnormalities of the dermatoglyphics: the crease-lines and the papillary ridges on palm and fingers, viewed as significant of mental deficiency. Only *the main abnormalities of the creases* have been listed: such as the presence of the simian and the absence of the long longitudinal line and faults in the two main transverse creases.

I want to repeat here what I mean by 'faults'. Most principal crease-lines, in either hands of normal or abnormal people, have some 'faults', that mean they are either slightly deviated in direction or show some interruption in their design. I have marked under the heading 'faults' only the more obvious deficiencies, such as abnormal position of the crease concerned, a number of 'isles' in the same line, very marked interruptions, doubling of the line or a particularly 'awkward' design. Also creases which are broken up into a number of pieces seem to be characteristic of 'abnormal' mental or psychological conditions. These kinds of 'faults' have been marked in this table.

SPECIAL FORMS OF THE PAPILLARY RIDGES

Whorls and loops on the thenar and hypothenar eminences are not usual in human hands, as I have pointed out earlier in this book. The most simple form of the papillary ridges, namely the arc-triangular one, also belongs to the unusual forms, and both specialities have been noted by other research workers as well as myself, to occur frequently in the hands of mental defectives.

The results of this table are exactly what one would expect.

SIMIAN LINE

The simian line occurred far more often in the mentally defective than in normal children, and

THE LONG LONGITUDINAL LINE

was most often absent in the high-grade boys.

Faults of the Crease-Lines were far more frequent in the defective groups of children and those of the lower transverse lines were predominant.

Needless to say that the more frequent presence of the simian and absence of the long longitudinal line are not surprising because of the predominance of the elementary regressive hand in mental defectives.

The faults in the design of the main creases have probably to do with abnormality and a general clumsiness of psycho-motor behaviour, which is reflected in the shape of the fingers and the creases of the hand.

Abnormal Forms of the Papillary Ridges occurred most often in the high-grade boys investigated, but the difference was not very great in regard to the same item in normal boys.

The difference between normal and high-grade girls was more marked in this respect, but it is interesting to note that abnormal patterns of papillary ridges on the palm were more frequent in normal boys than in high-grade girls. As these particular forms of the papillary ridges occur in a fairly large number of normal children, no diagnostic significance can be attached to them in regard to mental deficiency. This may, however, signify some kind of 'nervous' disturbance of a constitutional kind, common to persons both with and without mental deficiency.

INDIVIDUAL ASPECT

The second part of this chapter deals with individual cases.

I have tried to show by means of short diagnoses from hand-traits in 25 individual cases how the method of hand-interpretation, which I have expounded in this book, works in practice. The diagnostic statements are kept to the most essential characteristics. All the hands interpreted belong to high-grade boys and my statements were compared with the psychiatric diagnosis of the medical Superintendent.

The interest of this proceeding is obvious: first, it shows the kind of assessment that can be made from the hand and secondly it tests my method against other estimates.

Note: These diagnoses from hand-features were made independently from the psychiatric diagnoses. Nothing was known to me about any of the children investigated but their names and ages. No contact was made with them while they were under examination. No questions were asked so that any information should be drawn from the hands only. The investigation of the hands took in each case about 4 minutes and about 30 boys had their hands examined at a time.

TWENTY-FIVE HAND-INTERPRETATIONS

TOGETHER WITH THE DIAGNOSIS MADE BY THE PSYCHIATRIST

I have introduced every diagnostic statement by an indication of the most relevant hand-traits. It was of course impossible to go into a detailed description of all hand-traits which have led to my interpretations.

I have therefore described only the general hand-type and especially striking hand-features.

CASE 1

Age: 12 years 6 months.

Hand-type: elementary simple. Normally developed nails. The fifth finger is too short. Most simple pattern of creases without abnormalities.

HAND-DIAGNOSIS

Elementary simple hand. This is a normal boy except for a mild psycho-physical infantilism. Good chances of adaptation to a simple but normal life outside an Institution. He is pyknosomatic. His neurosis is mostly conditioned by environment.

PSYCHIATRIC DIAGNOSIS

This boy is not a mental defective. His I.Q. is 96. Suffers from a sexual neurosis with depression and guilt. Is suggestible. Has recovered quickly. He has probably witnessed sexual acts. His father is immoral.

CASE 2

Age: 13 years 3 months.

Hand-type: Long sensitive. Tapered finger-tips, normally developed nails. Overlong finger-tips. 'Fragile' fingers.

HAND-DIAGNOSIS

Asthenic boy; might be eunochoid. Dysplastic physique; possible psychosis. In any case a weak schizoid personality. Very much anxiety. No concentration and no capacity for integration.

PSYCHIATRIC DIAGNOSIS

Physically and sexually completely infantile. Highly anxious. Completely schizoid. Good verbal capacity. No concentration.

CASE 3

Age: 14 years 2 months.

Hand-type: mixture of fleshy and bony motor type. Very complex hand. Long nails, overlong finger-tips. Hands strong and broad. Very long fingers.

HAND-DIAGNOSIS

Intelligent boy (no question of mental deficiency) with strong vitality and dynamism, but a too strong inhibitory mechanism, revealing schizoid features of temperament. Very excitable with terrific inner tension and inadequate self-expression.

PSYCHIATRIC DIAGNOSIS

Infantile physique. No sign of pubic hair. Psychopathic personality. Very violent inner tension. Very intelligent, delinquent boy. I.Q. 100.

CASE 4

Age: 14 years.

Hand-type: elementary regressive with long fingers. Very small, almost rudimentary nails. Smooth palmar skin (unusual in this type of hand).

HAND-DIAGNOSIS

This is a mental defective, very sensitive and very inhibited boy. Extremely anxious. Submissive, and easily led.

PSYCHIATRIC DIAGNOSIS

Sensitive, grossly anxious mental defective. Very infantile emotionally.

CASE 5

Age: 15 years 7 months.

Hand-type: elementary regressive with long fingers. Spatulate tips; rudimentary nails; 'inarticulate' fingers, hyperflexible palm and finger-tips.

HAND-DIAGNOSIS

Infantilism with mental deficiency. Hypopituitary? Weak personality with weak drives and emotions. A schizoid of the dull variety. Very much anxiety but too scattered and not intelligent enough to develop real conflict.

PSYCHIATRIC DIAGNOSIS

A hypopituitary, pyknosome infantile boy. Passive homosexual. Little guilt. Very sensuous. Seduced other boys. Longs for caresses.

CASE 6

Age: 14 years 6 months.

Hand-type: elementary regressive hand of a mongoloid type. Fifth finger far too short, hand very broad with highly flexible palm.

HAND-DIAGNOSIS

A very high-grade mental defective with strong extravert tenden-

PSYCHIATRIC DIAGNOSIS

An infantile boy. Rich gesture. Good mimic. Athletoid physique

cies. He possesses great physical agility. Probably excellent at games of an 'acrobatic' kind. Emotionally unstable; affectionate; strong anxiety; highly emotional.

with infantile features. Very affectionate. Impulsive. I.Q. 80.

CASE 7

Age: 15 years.

Hand-type: sensitive small, acromicry. Length of hands: R.: 16·4 cm.; L.: 16·5 cm. Well-developed nails and tapered finger-tips.

HAND-DIAGNOSIS

A very abnormal boy; completely infantile; both physically and emotionally 'babyish'. Very lazy and lacking in concentration. Weak emotions; apathetic.

PSYCHIATRIC DIAGNOSIS

Completely infantile. No drive. Apathetic. Very 'nervous' and anxious. Masturbation.

CASE 8

Age: 14 years 8 months.

Hand-type: Elementary regressive with short fingers. Fifth finger deformed. Very 'primitive' hand; practically no accessory creases. Palms broad and very flexible. Arc-triangle pattern of papillary ridges on all finger-tips.

HAND-DIAGNOSIS

'Primary' personality with mental deficiency. Dull and kind-hearted. Very affectionate and 'warm'. Easily led.

PSYCHIATRIC DIAGNOSIS

Athletoid physique. Mentally defective, dull, heavy type. Submissive. Emotionally primitive; is very suggestible. Very 'good' boy. Fugues.

CASE 9

Age: 14 years.

Hand-type: Bony motor hand with simian line, which is an unusual combination. Many accessory crease-lines, which is in line with the bony motor type of hand. Spatulate fingers with rudimentary nails.

HAND-DIAGNOSIS

Leptosome, defective boy with very strong introvert tendencies, could be called a schizoid personality. Weak-minded and weak emotions. Apathetic.

PSYCHIATRIC DIAGNOSIS

Withdrawn, hungry of affection. Leptosome asthenic, not homosexual. Father schizophrenic. Comes from a good family.

CASE 10

Age: 14 years 1 month.

Hand-type: Long leptosome hand with mongoloid thumb and fifth finger. A very abnormal hand as the latter hand-traits are a rarity in this type of hand. Spatulate finger-tips and rudimentary nails. The hands are hyperflexible, another feature characteristic of 'mongolian' imbecility.

HAND-DIAGNOSIS

An asthenic and infantile mental defective, probably eunochoid. A 'babyish' boy. Sensitive, nervous and anxious. Feminine traits of personality. Submissive.

PSYCHIATRIC DIAGNOSIS

Homosexual, eunochoid boy. Very anxious and nervous. Has a father fixation. His father had left his mother.

CASE 11

Age: 15 years 4 months.

Hand-type: Elementary regressive with short fingers and over-developed bumps of sensitivity. Spatulate finger-tips with rudimentary nails, which are at the same time concave. Very flat palm.

HAND-DIAGNOSIS

'Primary' mental defective personality of a 'spastic' or epileptoid reaction type. Jumpy, impulsive, with marked introvert tendencies. Very strong inhibitive factor. Dull, cold and depressed person.

PSYCHIATRIC DIAGNOSIS

Mild infantilism. Very jerky. Stammers. He is dull and withdrawn. Sulky. Very difficult to handle.

CASE 12

Age: 13 years 9 months.

Hand-type: Unusual hands. Far too long and large, of the elementary simple type. Well-developed nails. No sign of mental deficiency.

HAND-DIAGNOSIS

This boy is physically very strong and precocious. He is a highly excitable person, very suggestible. Probably suffering from an anxiety neurosis of the hysterical reaction type.

PSYCHIATRIC DIAGNOSIS

Overmature. Leptosome, well-built boy. Anxious, neurotic bed-wetter. Compulsive trends. Well behaved. Fallen behind in school. Not defective. Did not go to school.

CASE 13

Age: 14 years 5 months.

Hand-type: elementary regressive with long fingers. Very large and 'inarticulate' thumbs; simian line in right hand; powerful thenar eminences. Under-developed nails, but not rudimentary.

HAND-DIAGNOSIS

Impulsive and jerky. Very strong introvert tendencies; moody. Physically active and probably a good labourer. Emotionally weak and very dependent on others.

PSYCHIATRIC DIAGNOSIS

A schizoid personality. Infantile; very sensitive. Jerky and moody. Good worker. Loves animals.

CASE 14

Age: 14 years 6 months.

Hand-type: elementary regressive narrow hand, with a short and pointed thumb and a likewise very short fifth finger. No other abnormalities.

HAND-DIAGNOSIS

Asthenic boy, lazy, with feminine tendencies, probably homosexual, definitely infantile. Submissive but probably impulsive.

PSYCHIATRIC DIAGNOSIS

Leptoasthenic type. Sexually absolutely infantile. Homosexual. Suggestible; extremely impulsive. Hysterical. Unstable personality. Anxiety. High I.Q.

CASE 15

Age: 15 years 5 months.

Hand-type: very small elementary regressive hands, narrow, with a misshapen fifth finger. Spatulate finger-tips and rudimentary nails.

HAND-DIAGNOSIS

Slightly defective, infantile boy with strong instincts and drives. Affectionate and emotionally very dependent.

PSYCHIATRIC DIAGNOSIS

A little dwarfed. Pyknosomatic. Affectionate, self-dependent. Rich phantasy life. Rather moody.

CASE 16

Age: 14 years 6 months.

Hand-type: Very small hands, almost amounting to acromicry. Narrow hands with particularly short fifth finger. Very faulty creases.

HAND-DIAGNOSIS

Physical and mental debility. Infantile, asthenic boy; schizoid. His drives and emotions are weak. Neurasthenic, anxious. Very inhibited, but suggestible. He always follows and never leads.

PSYCHIATRIC DIAGNOSIS

Real mental defective. Very weak personality and highly suggestible. Preoccupied with sex.

CASE 17

Age: 14 years 7 months.

Hand-type: Very large and very broad elementary regressive hands with too long a thumb and too short a fifth finger. Spatulate finger-tips and rudimentary nails. Simian-like line.

HAND-DIAGNOSIS

Mental defective but physically very strong boy. Very well developed. He is a depressed person with much anxiety. Schizoid and impulsive. A very difficult child.

PSYCHIATRIC DIAGNOSIS

Athletoid physique. Sexually mature—schizoid. Will probably develop catatonic schizophrenia. Tried to kill a schoolteacher. Motor-behaviour awkward.

CASE 18

Age: 15 years.

Hand-type: dysplastic, most abnormal hand of all the high-grade boys investigated. Length of hands: R.: 15·6 cm. and L.: 16·0 cm. These are acromicric hands but there are more abnormalities. Ankylosis of fifth finger in the right hand, very big bumps of sensitivity on all fingers, tapered finger-tips with rudimentary nails; and a rigid thumb as if made of 'one' piece (this is a characteristic trait in the lower grades of mental deficiency).

HAND-DIAGNOSIS

This is a hand revealing both advanced mental deficiency and probably psychosis. If the latter does not exist, the personality of this boy is so under-developed and weak, that one cannot speak of a personality at all.

It surprised me to find such hands among the high-grade group.

PSYCHIATRIC DIAGNOSIS

Completely infantile boy. Probably pre-psychotic. I.Q. 57.

CASE 19

Age: 15 years 6 months.

Hand-type: flat leptosome or bony motor hand, with long fingers and well-developed nails, and spatulate finger-tips.

HAND-DIAGNOSIS

Leptosome, neurasthenic boy. Very strong introvert tendencies and very strong inhibitive factor accordingly. Highly anxious and not mentally defective enough to avoid feelings of guilt and conflict. Very easily suggestible.

PSYCHIATRIC DIAGNOSIS

Leptosome epileptic. Hypochondriac. Very anxious. I.Q. 78.

CASE 20

Age: 14 years 2 months.

Hand-type: elementary regressive with short fingers, recalling the typical 'mongolian' hand, with a short clubbed thumb and a very short fifth finger. The simian line is also present. The nails are well developed.

HAND-DIAGNOSIS

An under-developed, infantile personality with marked extravert tendencies. Uncontrolled emotions, very affectionate and kind-hearted. In this way easily to be led. May be up to all sorts of tricks, but never mischievous.

PSYCHIATRIC DIAGNOSIS

Infantile physique. Very suggestible. Highly emotional. Always in a state of excitement or on the verge of tears. Plucky and well behaved.

CASE 21

Age: 13 years 5 months.

Hand-type: This is an 'endocrine' hand, difficult to class among the principal hand-types. The palm is fleshy and broad, the fingers are very long and tapered; the palmar skin is very fine and the creases superficial. A good many accessory crease-lines.

HAND-DIAGNOSIS

This boy is an 'endocrine' case, of a hypopituitary dysfunction. He is inhibited to such a degree that one can speak of a schizoid personality. A moody child, very sensitive and touchy. Very marked anxiety; he is probably capable of much conflict.

PSYCHIATRIC DIAGNOSIS

Hypopituitary mental defective. Very little emotional expression, but he thinks a good deal. Anxiety and conflict, but no 'gross' anxiety. Guilty about masturbation.

CASE 22

Age: 11 years 11 months.

Hand-type: small sensitive with spatulate fingers and normally developed nails. Very abnormal position of the upper transverse line. Overlong finger-tips.

HAND-DIAGNOSIS

Very slight mental deficiency or none at all. This is a highly sensitive, physically very under-developed child whom I suspect to be pre-psychotic.

PSYCHIATRIC DIAGNOSIS

Infantile asthenic with structural instability. Self-centred, suggestible and of normal intelligence. Probably psychotic.

CASE 23

Age: 13 years 7 months.

Hand-type: Unusual hand, difficult to class. The shape corresponds most closely to the sensitive small hand, the crease-lines more to the elementary regressive pattern. Simian line. Over-developed bumps of sensitivity, tapered finger-tips.

HAND-DIAGNOSIS

An infantile psychopath, with strong inhibitions and very marked introvert tendencies. A weak personality without physical or emotional dynamism.

PSYCHIATRIC DIAGNOSIS

Mildly defective boy. Very inhibited, schizothyme. Had behaved badly.

CASE 24

Age: 12 years.

Hand-type: Long sensitive with very abnormal fingers: the thumb and fifth finger are minute, which is most uncommon in this type of hand. Tapered finger-tips, broad and normally developed nails. The hands are very oversized, measuring 18·0 and 18·5 cm. and very broad.

HAND-DIAGNOSIS

Very little or no mental deficiency. An endocrine case; this boy is eunochoid. Marked feminine traits of personality. Very infantile, impulsive and erratic. Highly suggestible. Probably passive homosexual.

PSYCHIATRIC DIAGNOSIS

Pyknosomatic physique. Eunochoid. Well behaved at the level of a baby. Flirts with other boys. Passive homosexual.

CASE 25

Age: 15 years.

Hand-type: Narrow and small elementary regressive hands with long fingers. Spatulate finger-tips and very small nails. Clubbed thumbs.

HAND-DIAGNOSIS	*PSYCHIATRIC DIAGNOSIS*
Mental defective of the rather typical endocrine variety. He is hypopituitary, of the kind described as Levy Lorraine syndrome. Infantile, neurasthenic boy with marked introvert tendencies. Not grossly defective. Suggestible and anxious. Submissive and easily led.	Leptosome-asthenic boy. Poorly developed. Anxious, suggestible, excitable. Easily becomes unhappy. Goes with anybody, does not run away alone.

Apart from abnormal cases, I made detailed hand-interpretations of 19 schoolchildren, all girls of normal intelligence. My reports on them were compared with the estimates of several teachers. These have been written up by one of them, who gives an account of the experiment and its results.

In February, 1944, nineteen girls, all aged about 11 years and just about to enter a well-known London school, submitted their hands for examination and, with the parents' permission, allowed prints to be taken in the normal way. The estimates formed by Dr. Wolff of the physical constitution, temperament and general ability of these children were all locked away without anyone seeing them for a year, at the end of which time opinions on these same children were collected from the various mistresses teaching them. The papers were compared and then again locked up. Four years later still, in December 1949, without reference to any of the earlier papers, estimates from the Head Mistress and staff were once more collected. By this time there was every reason for feeling that the children were really well known. In the following pages a general summary of the teachers' estimates is set alongside the original brief outlines given by Dr. Wolff from the hand-interpretations only, and a comparison indicated.

ESTIMATES FROM HAND ONLY. 1944	*SUMMARY OF ESTIMATES FROM TEACHERS. 1945 AND 1949*
CASE I	
Hyperthyroidism (?tubercular). Nervously very delicate. A hypersensitive and 'introverted' child. Excitable and restless with marked anxiety. Intelligent and imaginative child with quick reactions.	A rather undersized child—physically a late developer. Nervous and excitable. Average mentality. Slightly unco-operative and sometimes seemed lethargic. Family not very helpful.

COMMENT

The agreement is good. A strongly introverted child who is also anxious may well appear unco-operative or lethargic.

ESTIMATES FROM HAND ONLY. 1944	*SUMMARY OF ESTIMATES FROM TEACHERS. 1945 AND 1949*
CASE 2	
A delicate and neurasthenic child—physically backward but mentally advanced. Highly sensitive and excitable (all over the place). Strong overcompensation of her delicacy through applying herself intellectually. A very 'mercurial' child with quick flexible reactions. Extremely aware of surroundings. Very intelligent. Imaginative but dispersed. Versatile.	An undersized child, brought up by her grandparents. An extravert, anxious to please, of the quick sensitive type with considerable ability but little concentration. Very all round. She developed well, in spite of bad war experiences and being often over-fatigued. The child of an unmarried mother. Has become reliable, cheerful, making good contacts.

COMMENT

Good agreement. This child now promises well as a potential teacher of gymnastics and seems to be happily adjusted.

CASE 3	
Health impaired through rheumatic tendency and probably circulatory weakness. Very inhibited with marked anxiety. Easily depressed. Not easily affectionate but having once taken to a person will not alter. Nervous weakness but a sense of purpose. Reactions of a neurasthenic type although she shows will power and determination to overcome handicaps. Slow in verbal expression but with some structural sense and intelligence. Possibly good with hands.	A tall, awkward child, lacking in confidence. Often with good ideas behind a muddled expression. Reliable. Works willingly for those in whom she has confidence, but often antagonistic and then obstinate. Had terrible war experiences. Generally stiff and unco-ordinated. Would persevere if convinced a thing was worth doing. At Butlin's Camps was a table tennis champion. Otherwise clumsy with her hands. Wrote very badly. Learned the piano.

COMMENT

Complete agreement, except perhaps over use of hands—but table tennis requires both hands and will power for co-ordination. Also, though failing general School Examination in most subjects, she passed in Domestic Science with a practical test.

ESTIMATES FROM HAND ONLY. 1944	*SUMMARY OF ESTIMATES FROM TEACHERS. 1945 AND 1949*
CASE 4	
A very strong child physically, with good vitality. An extremely well-balanced and strong personality. Kind-hearted, affectionate and placid. Great natural integrity. Not intellectual but intelligent and practical. Structural more than verbal. Perhaps musical?	Healthily normal child in all ways. Had bad war experiences and lived in over-crowded conditions. Grew into a large comfortable person, quiet and hard-working. Perhaps a little reserved and diffident. Good average mentality.

COMMENT

The agreement here is obvious, the more so now that this girl has grown into a useful and responsible senior in the school and has chosen to enter the nursing profession.

CASE 5	
A strong bony child with good vitality and dynamism. Vivacious but emotionally stable. Self-possessed and predominantly introverted. Keeps to herself and perhaps had a tendency to be too reserved for her age. Might be good at outdoor sport (such as tennis?). Clever with her hands. 'Intellectual type' with a structural sense and synthetic views.	A reserved and slightly unresponsive child with very good ability. She developed precociously and became angular, physically and in other ways. She became unco-operative and individualistic and did not trouble to use her good natural ability. She was the youngest of a much older family and was not happy at home. She was ill at intervals and not disposed to join well in team games.

COMMENT

Good agreement. This girl seems happier now that she has left school and begun a business training.

ESTIMATES FROM HAND ONLY. 1944	*SUMMARY OF ESTIMATES FROM TEACHERS. 1945 AND 1949*

CASE 6

Leptosome with good vitality. Graceful in movement and may be good at games. A well-poised and emotionally balanced child, predominantly introverted. Not too forthcoming and might be considered older than her age. A sense of purpose and determination.

A promising child at first, healthy and normal in her reactions, with both practical and aesthetic interests, sociable, normally self-confident and co-operative. A good leader.

Later she became a fairly courteous but determined rebel, did the minimum of work and appeared far too mature outwardly for her age.

Her father died tragically when she was about 14. She told no one. Her mother looked the 'glamour girl' and was un-motherly to her.

COMMENT

There is no doubt that in this case intellectual promise was unfulfilled. The child altered noticeably between 13 and 14. This was possibly due to the mother's attitude.

CASE 7

Very strong vitality and physique and very healthy. Should be good at more exacting sports. Simple, rather primitive make-up. Shy and affectionate with strong instincts and straightforward reactions. Emotionally stable with a maternal sense. Apt to protect the younger or weaker.

A fair, healthy looking child, with good physique, easy to deal with; interested in games and practical things.

Though very willing to work, she had no natural gifts and no idea of her own shortcomings. She was determined, capable in some ways and reliable and became a good leader in her form.

COMMENT

This very straightforward case shows excellent agreement.

CASE 8

Physically very strong child but with slight hypothyroidism—her

A good healthy child capable but not persistent over her work. She

ESTIMATES FROM HAND ONLY. 1944

nervous health less good than her organic condition. Rather moody but extremely kind-hearted, affectionate and forthcoming. Very easily led. She has will-power but little nervous solidity so there is a terrain of anxiety. 'Grown-up' mentality with fine perception and possibly fine expression (? musical). Imaginative child with a natural taste for beauty and a love of comfort.

SUMMARY OF ESTIMATES FROM TEACHERS. 1945 AND 1949

grew into a tall nervous girl, self-conscious, with some slight hesitation in speech. She seemed apprehensive but she managed her size well. She had no help at home (was sometimes beaten by her father). She was pleasant and easy to deal with.

She left school early owing to difficult home circumstances (her father disappeared). She had a maternal attitude towards younger children.

COMMENT

The agreement is obvious. But the 'fine expression' of which probably this child was capable had little chance of developing with such a background.

CASE 9

A delicate child with little nervous resistance. 'Introvert' type. Sensitive and shy. Might easily be on the defensive as a general attitude. In spite of inner tenseness is good-willed and obedient.

Mentality — (indirect conclusions). Children of such constitution often show a good intelligence and, as physical powers fail them, develop as a compensatory measure their imaginative capacities.

A retiring child, lacking in confidence. Not forthcoming, easily discouraged. Interests perhaps mainly aesthetic.

A fair-haired petite person—quiet and uncommunicative—not without ability but tiring easily. An only child and rather spoiled. A late developer, keeping herself aloof.

She showed some artistic taste and became good with her hands.

COMMENT

The agreement is obvious. This girl afterwards chose to train as a hairdresser.

CASE 10

A strong child with slight tendency to hypothyroidism. Likely to be good at some sports. An extra-

A strong child from a difficult home with crowded living conditions. Not academic; better at

ESTIMATES FROM HAND ONLY. 1944

vert, interested in her surroundings. A rich emotional life but moody and easily 'dispersed' with traits of instability.

A practical child—not intellectual, with manual and physical gifts—possibly musical.

SUMMARY OF ESTIMATES FROM TEACHERS. 1945 AND 1949

practical things and games. Lacking in persistence and careless. Appeared to be willing but would actually escape responsibility. Later was ill with some heart trouble which stopped her games. She showed less and less interest in school and had no home backing, so left before facing any examination.

COMMENT

The correspondence here, too, is striking. Since leaving school she has been in and out of jobs. The background difficulties reacted too heavily on a child so easily dispersed and unstable.

CASE 11

This child's hands are the most abnormal of the group (simian line). She is a powerful, well-developed child. The temperament is erratic and unbalanced—highly emotional with lack of control. Discipline and adaptation difficult to attain and unstable.

An 'introvert' of the adventurous type with physical courage.

A good child with intellectual interests. Slightly lacking in confidence. Young for her age group. Under-developed while overgrown.

Much under the control of an over-careful mother.

Though naturally shy and self-conscious, she plays a prominent part. Though obviously smouldering under criticism would always control herself.

She has linguistic gifts. Has become a reliable, hard-working student and a good prefect.

COMMENT

This child is still young and her emotional development not far advanced. Here it seems that the possibilities suggested in the hand analysis may already have shown more at home than at school, and will probably show still more in after school life.

CASE 12

A strong, well-developed child with a tendency to hypopituitarism

A careful, conscientious child, working well and interested on the

ESTIMATES FROM HAND ONLY. 1944	*SUMMARY OF ESTIMATES FROM TEACHERS. 1945 AND 1949*
—will get stout. Lazy and phlegmatic in temperament, with superficial emotions and a certain escapism. A mixed type with introversion more marked than extraversion. Such people are often musical with good voices. Only fairly intelligent.	intellectual and aesthetic sides. Very reliable and helpful but a child with marked mannerisms — introspective. On the English side said to have considerable ability; in other subjects only average. Has no father and is probably compensating for this. Musical with a good voice. This is the child of an unmarried mother, with difficult home circumstances.

COMMENT

The agreement here is, on the whole, obvious, but the intellectual ability appears to be higher than the hand suggests. Has the escapism in her case found an outlet in what she writes? Is her intellectualism compensatory in origin?

CASE 13

Asthenic physique — highly strung. Sensitive with aesthetic needs and an aesthetic sense. A difficult emotional life. Lack of real contact with others. Eccentricity.	A tall, fragile-looking girl, who left school early owing to home circumstances. Difficult to know. Tastes seemed mainly practical or aesthetic. Rather self-centred and diffident. Not very capable or very persistent.

COMMENT

Agreement as far as the record goes. She left two years after coming to the school.

CASE 14

Physique or the leptosome variety. Nervous resistance low. Predominantly 'introverted' child but sensitive to her surroundings. Not easily roused emotionally but inclined to be faithful to those for whom she forms an affection. Possibly gifted for light sports.	A tall, thin child, quiet, self-centred and rather timid. Very conscientious worker. Interests intellectual and aesthetic. She was modest and unassuming with considerable ability and power of concentration. This child was overshadowed at home by an elder

ESTIMATES FROM HAND ONLY. 1944	*SUMMARY OF ESTIMATES FROM TEACHERS. 1945 AND 1949*
Might be good with her hands at some aesthetic pursuit (fine needlework or piano).	sister and nearly always over-fatigued. Frequently absent. When tired, lacked control.

COMMENT

The two pictures are of the same child.

CASE 15

A strong child with good dynamism. Nervous resistance impaired constitutionally through degenerative traits. A tendency to anxiety neurosis in a physically and mentally active child. Emotions strong and rich but she needs assurance and encouragement in her relationships. Probably good at sports and clever with hands.	A large, cheerful child, not much intellectual ability. Good at sports and on the playing fields a leader. During the war had some very bad shocks—deaths in the house. Lived for long in a shelter. Left in 1947 through home circumstances and being unable to stay the course.

COMMENT

The accounts agree.

CASE 16

Strong and graceful girl with physical dynamism. A marked kinaesthetic type; very self-centred (narcissitic tendencies). Low adaptability. Emotional life on a limited scale without proper objectivation and attachments. She wants contacts mainly to be admired and for this purpose develops 'charm' in order to please. Has an aesthetic sense in everything concerning movement—might be gifted for dancing or at graceful sports.	A tall, rather pretty child who was at first spoiled at home. Later the attitude of the parents abruptly changed and became over-strict. A child with very little mental ability who 'fancies herself'. Not willing to take responsibility. Generally weak but quite attractive and pleasant. Has some feeling for the lighter type of movement.

COMMENT

The agreement is complete.

ESTIMATES FROM HAND ONLY. 1944	*SUMMARY OF ESTIMATES FROM TEACHERS. 1945 AND 1949*

CASE 17

Delicate child with tendency to arthritis. Has great sense of purpose and staying power. Emotionally strong with good adaptability. Predominantly 'introverted', probably because of, or at least encouraged by, some physical delicacy. No special aptitudes can be diagnosed.	An intelligent child, thin and delicate looking, with wide interests. She showed in the gym some odd want of co-ordination and not until later was it found that she had had an iron on one leg before coming to the school. A capable girl who can make herself felt, with an adult view point. She made good contacts, was friendly, sensible and reliable. A good organizer.

COMMENT

An extraordinary agreement, especially about the physical delicacy mentioned above.

CASE 18

A well-developed child with good physical strength and dynamism. Sensitive to surroundings. Very responsive and adaptable. A placid child. An 'extravert' with love for outdoor activities and for nature. A maternal disposition. Applies herself not only well to people but to given tasks. Responsive and reliable. Probably good at games. Might be musical?	A well-developed healthy child who grew up quickly. Had good manual dexterity but became slack and uninterested in school—did very little work. She came from an unhappy home where parents quarrelled bitterly. Her loyalties were always divided, with no energy over for work. She was disgusted with herself when unable to succeed and took it out of others.

COMMENT

The disagreement between these two accounts of the same child is a sad commentary on the embittering effect of the divided home on a child 'sensitive to her surroundings and responsive'.

CASE 19

A strong child with good physical dynamism. A simple elemen-	A well-developed child with no intellectual or aesthetic interests and

ESTIMATES FROM HAND ONLY. 1944	*SUMMARY OF ESTIMATES FROM TEACHERS. 1945 AND 1949*
tary personality inclined to be materialistic, with too little capacity for sublimation. Emotionally dull but a tendency to be impulsive. Gifts on the purely physical side—gardening or swimming.	very indifferent home. Good physical ability—always friendly and pleasant. Quite content to be behind the rest of the class in most of the work and to muddle along uncertain of anything. A poor mentality. Left before taking a leaving examination.

COMMENT

A perfect agreement.

GENERAL COMMENT

The Diagnoses of Personality from the hand suggest possibilities which in no case can be completely fulfilled by the growing child while still at school. There are potentialities in every one of these cases which, in the nature of things, cannot be seen by the teachers during the life at a day school. The homes would have other things to tell and the later lives of these children will reveal far more of the ways in which their latent powers react with circumstances and develop with time. But all the same, in these 19 cases there are only two not showing remarkable agreement between the hand-interpretation and the teachers' estimates. These are numbers 6 and 18, and in these two the want of fulfilment of the latent power is marked. It is interesting to observe that in both these cases the home conditions were against the happy development of the child. On the other hand, the abnormality of the child with the simian line—number 11—was not specially observable at school and in that case there was careful home training, which no doubt inculcated some habit of control, sufficient to mask to a great extent the deeper emotional disturbances, though not entirely.

The other cases all show an amazing correspondence.

LUCY PRESTON

I have concluded this book by giving hand-diagnoses of these normal children, in order to give the reader an idea of how my work could be applied one day in ordinary life. Research on the 'abnormal', which figures so largely in this book, had to come first, in order to prepare the way for research on the 'normal'. Looked at from a broad and practical point of view, psychological diagnosis seems more important for the so-called 'normal' members of the community than for the mentally deranged, as the continuation of the physical and cultural existence of mankind depends on the former. This points me to the next step in my research. I must now test out my method on large groups of so-called 'normal' people.

1 Elementary simple type of hand

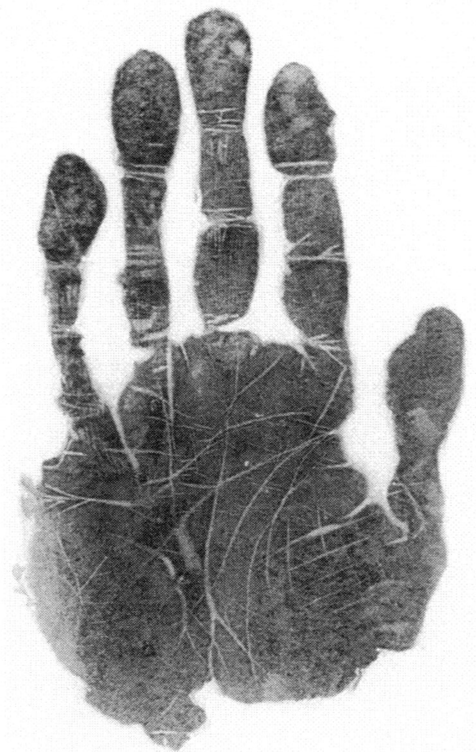

2 Fleshy motor hand

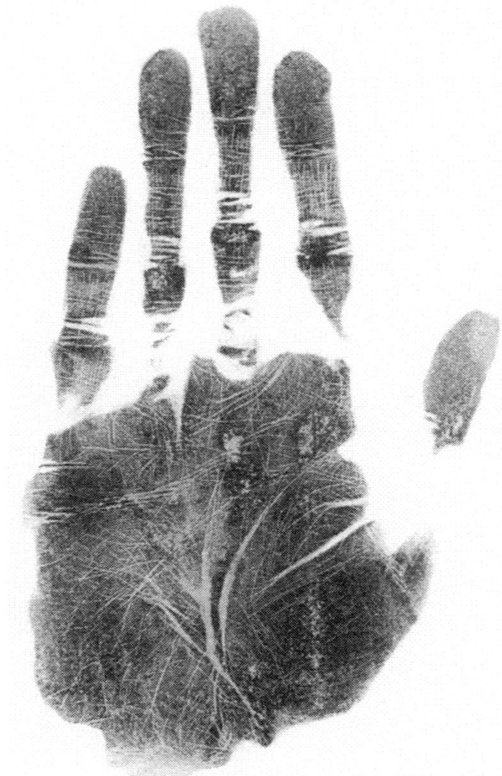

3 Bony motor hand

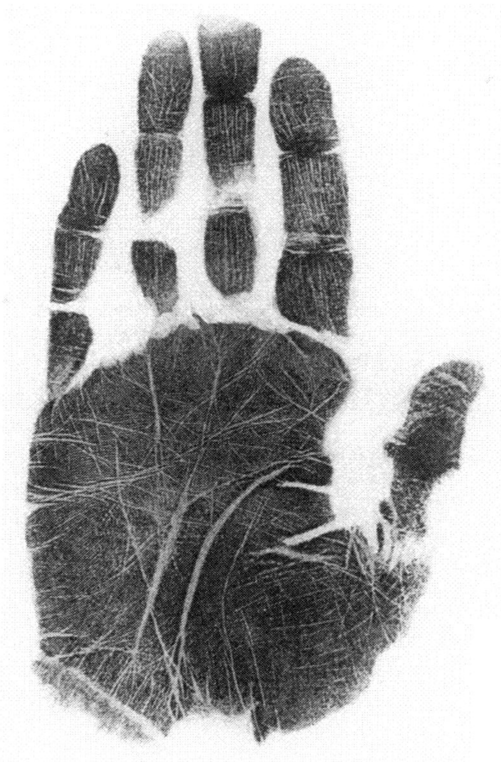

4 Sensitive small hand

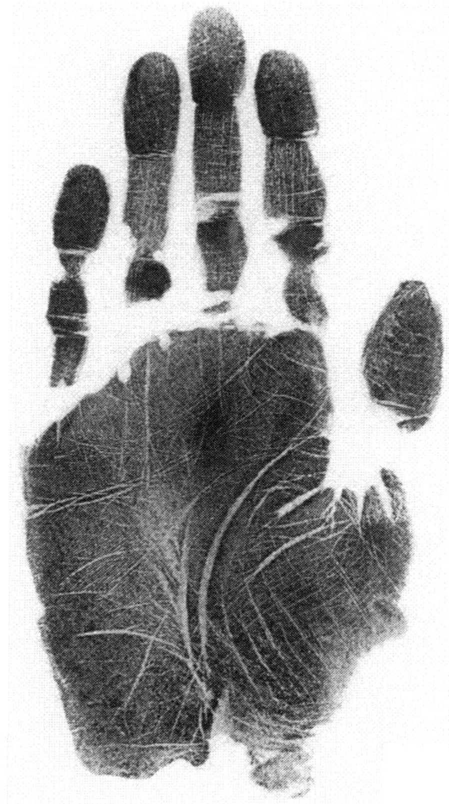

5 Sensitive long hand

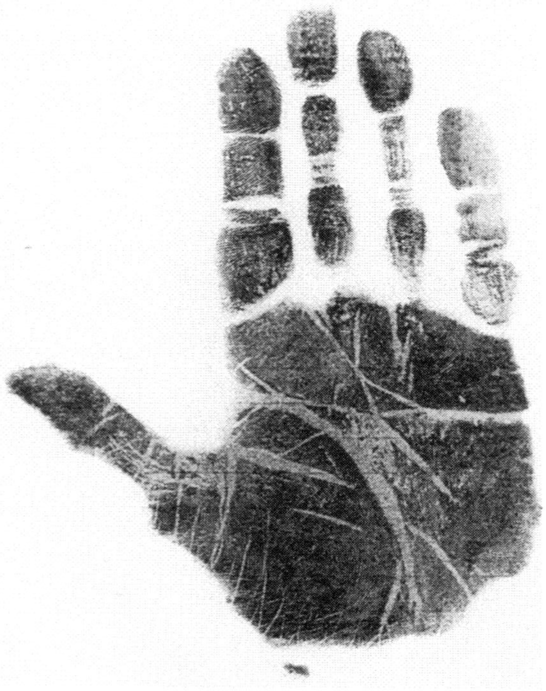

6 Elementary regressive hand of a moron

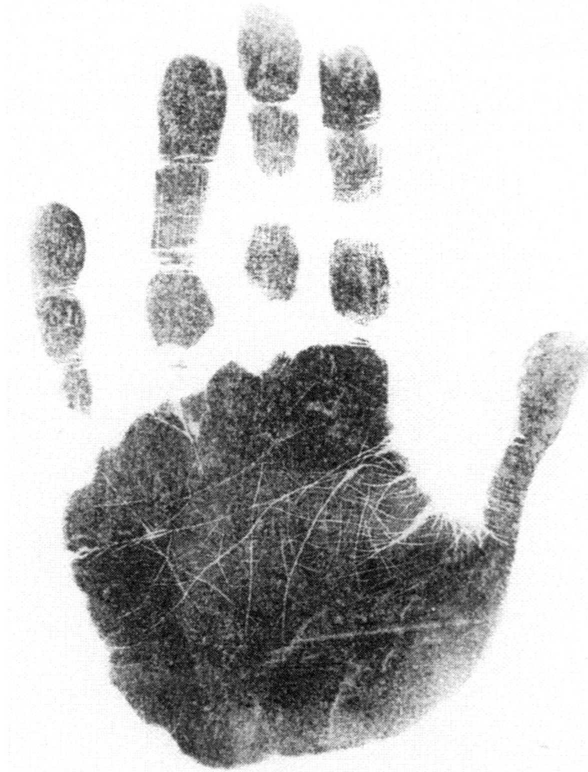

7 Elementary regressive hand of a very intelligent woman — with an eccentric personality

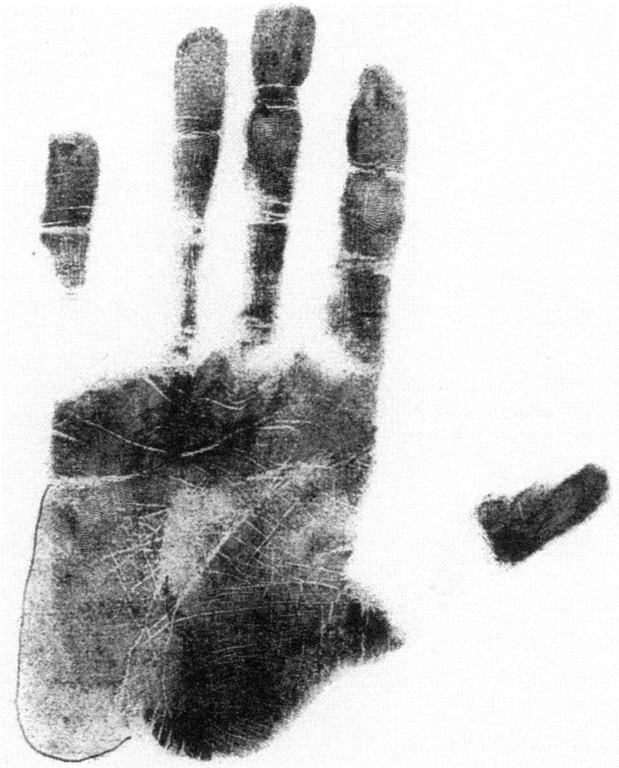

8 Hand-print of a eunuchoid man

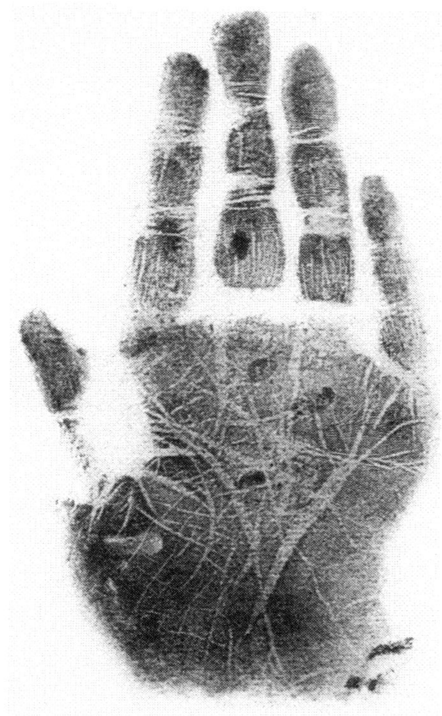

9 Hand-print of a eunuchoid boy of 12 years of age

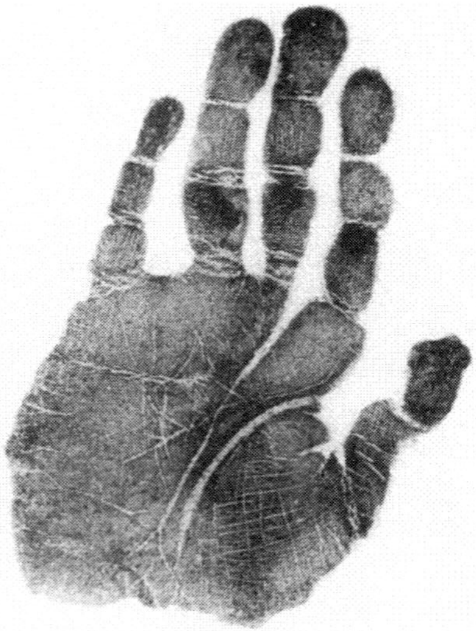

10 Hand-print of a cretin

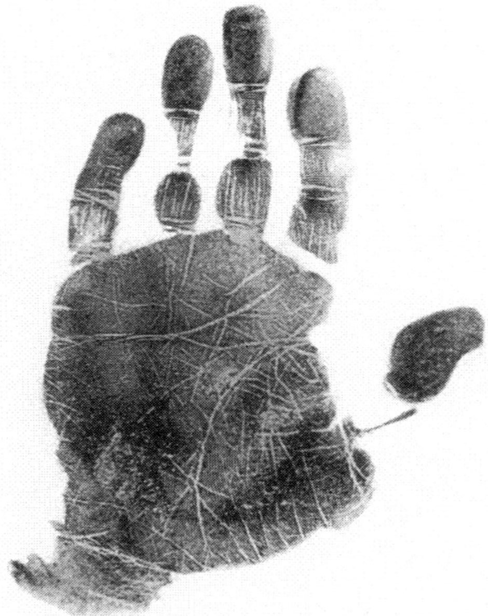

11 Hand-print of a Mongolian imbecile. Note the 5th finger – one crease on the minimal digit is missing

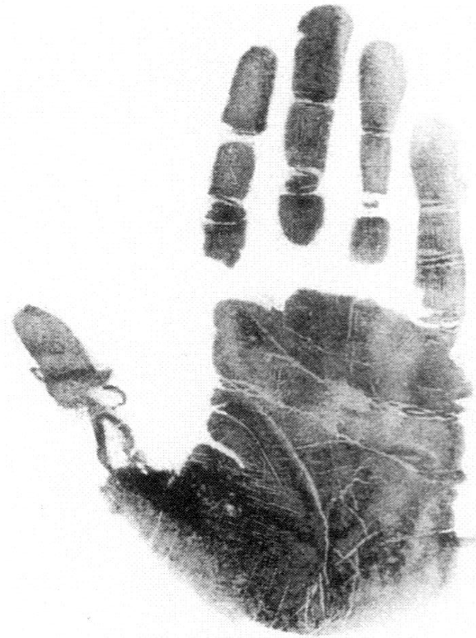

12 Hand-print of a high grade defective boy with Levy-Lorraine syndrome

HAND-PRINTS OF TWO NEUROTIC PERSONALITIES

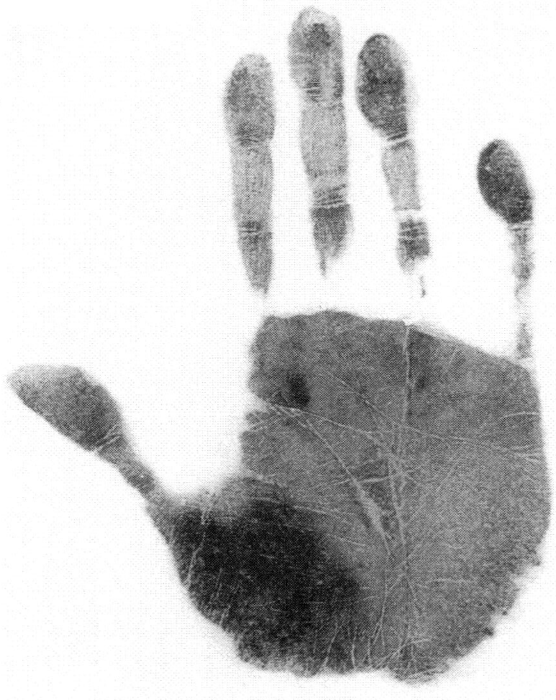

13 Of a young man of normal intelligence with marked disintegration of personality

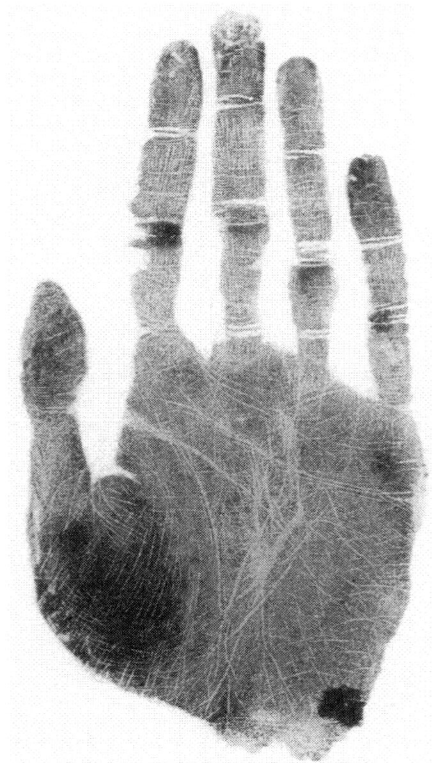

14 Of a young man of unusual intelligence with abnormally strong introvert tendencies, suffering from anxiety neurosis of an obsessional type

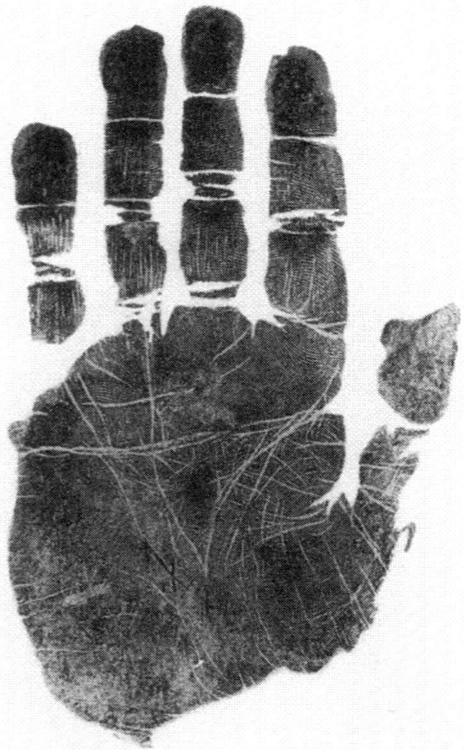

15 Male schizophrenic – Schizophrenia Simplex

16 Male schizophrenic – Schizophrenia Simplex

HANDS OF SCHIZOPHRENICS

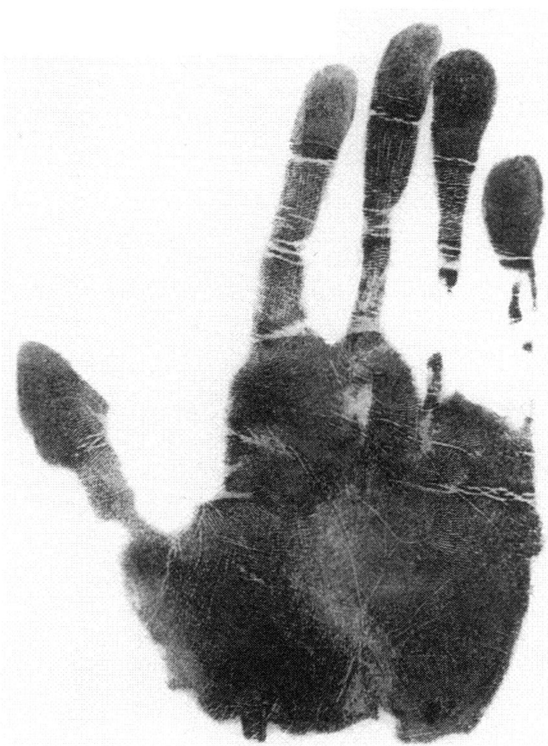

17 Hand-print of a male schizophrenic, suffering from the catatonic form of schizophrenia

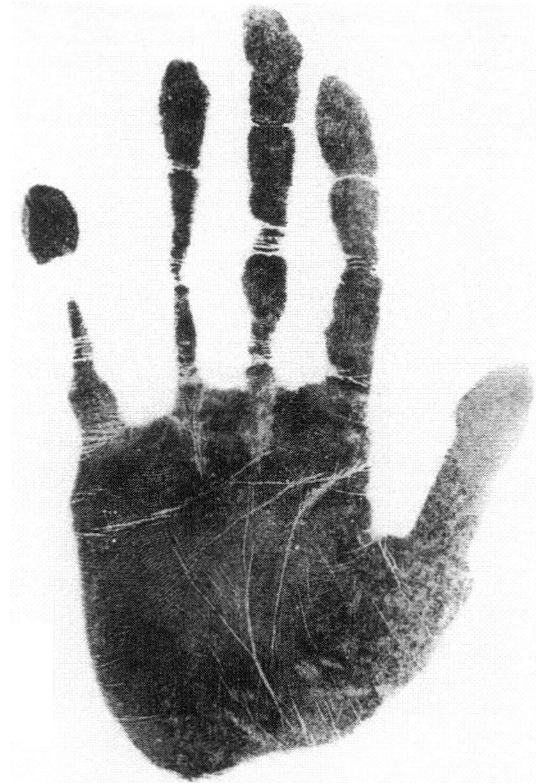

18 Hand-print of a woman schizophrenic, suffering from the catatonic type of the illness.
Note the extraordinary length of the fingers

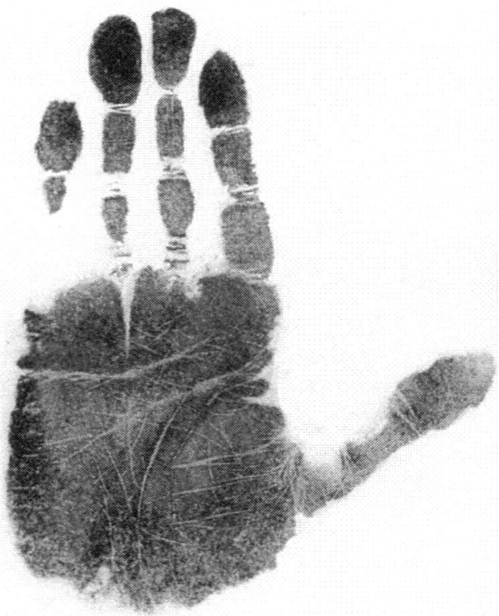

19 Female schizophrenic, suffering from the paranoidal type of the illness.
Note the almost equal length of the 3rd and 4th fingers

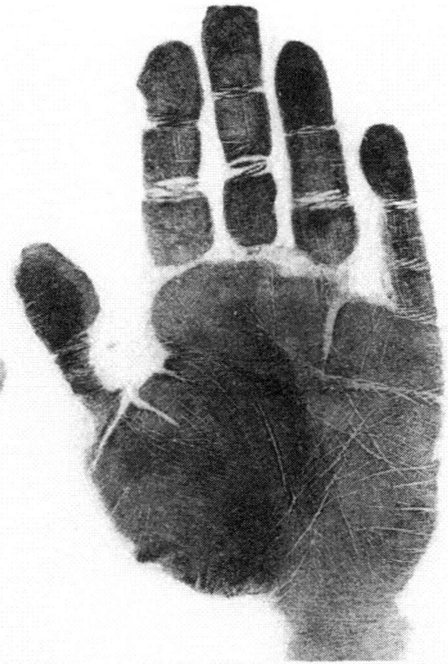

20 Female schizophrenic, suffering from the hebephrenic type of the illness

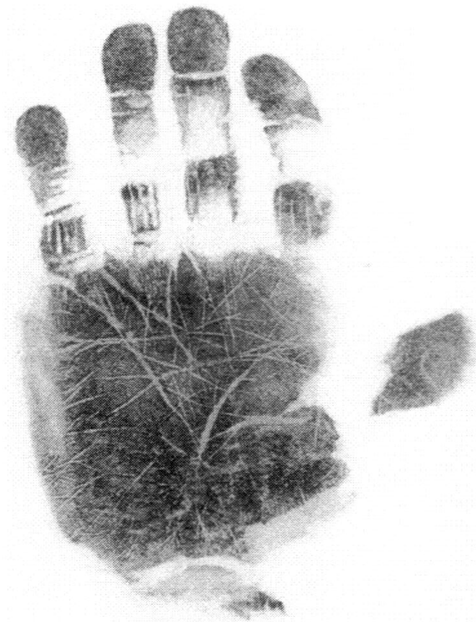

21 Woman with secondary dementia.
Note the typical hand traits of mental deficiency

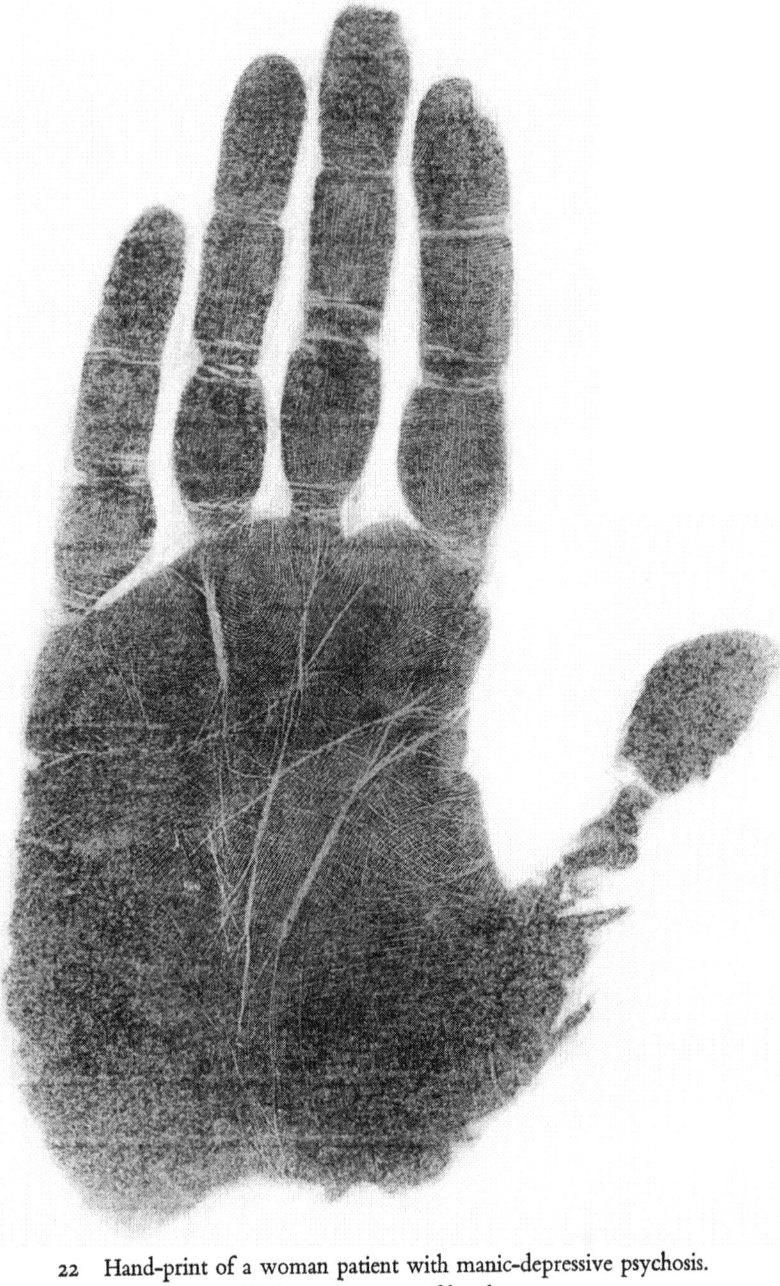

22 Hand-print of a woman patient with manic-depressive psychosis. Fleshy motor type of hand

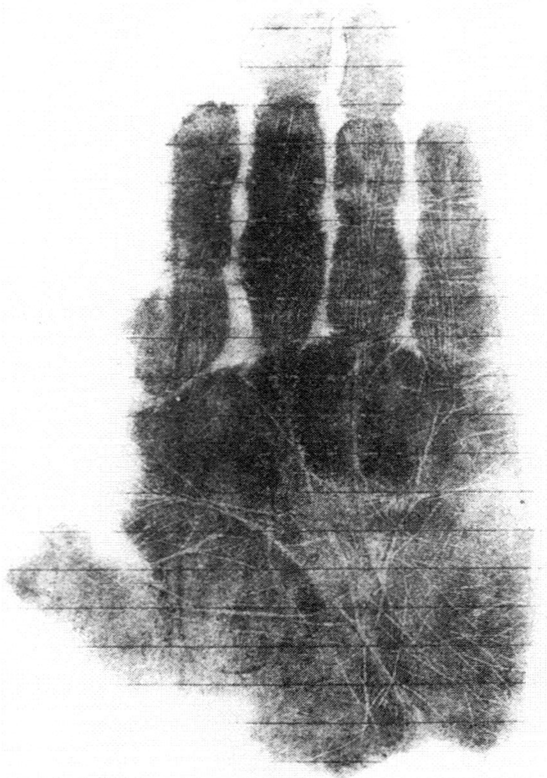

23 Hand-print of a woman suffering from paraphrenia.
Note the abnormal digital formula

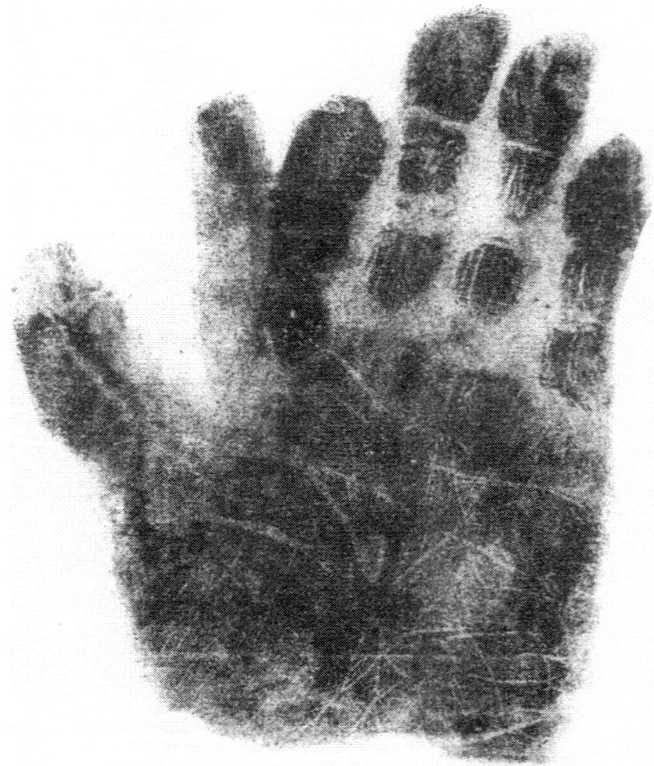

24 Hand-print of a male epileptic.
Note the abnormal digital formula – the 5th finger is longer than the index finger

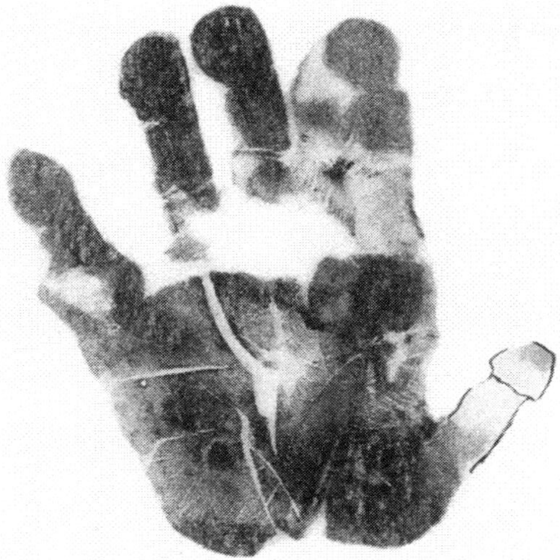

25 Hand-print of a female idiot, recalling that of a gorilla.
Note the simian line, the ape-like shape and simian insertion of the thumb

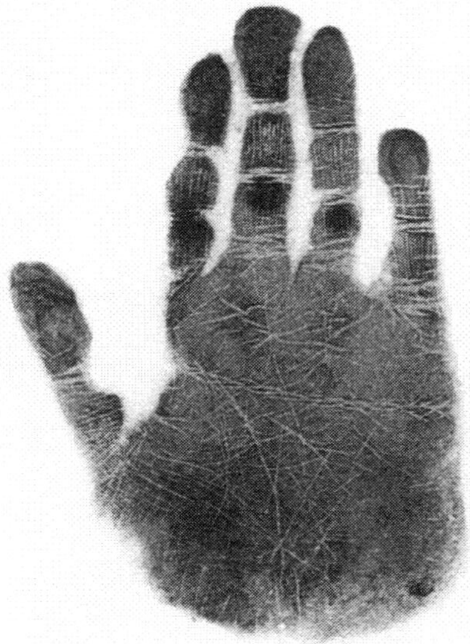

26 Hand-print of a low grade imbecile.
Note particularly the 5th finger where one crease is missing

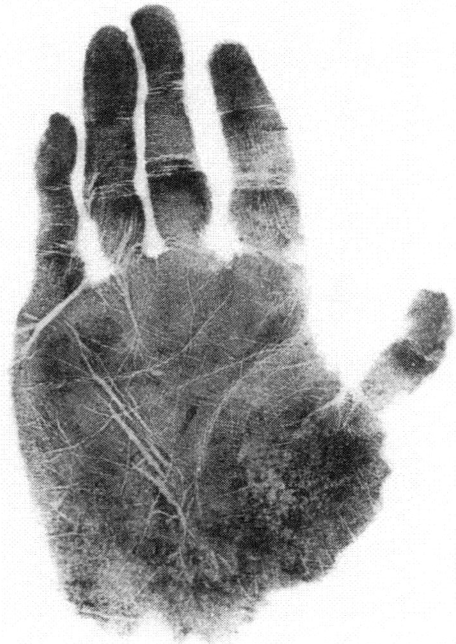

27 Hand-print showing a case of Arachnodactyly in a female low grade imbecile

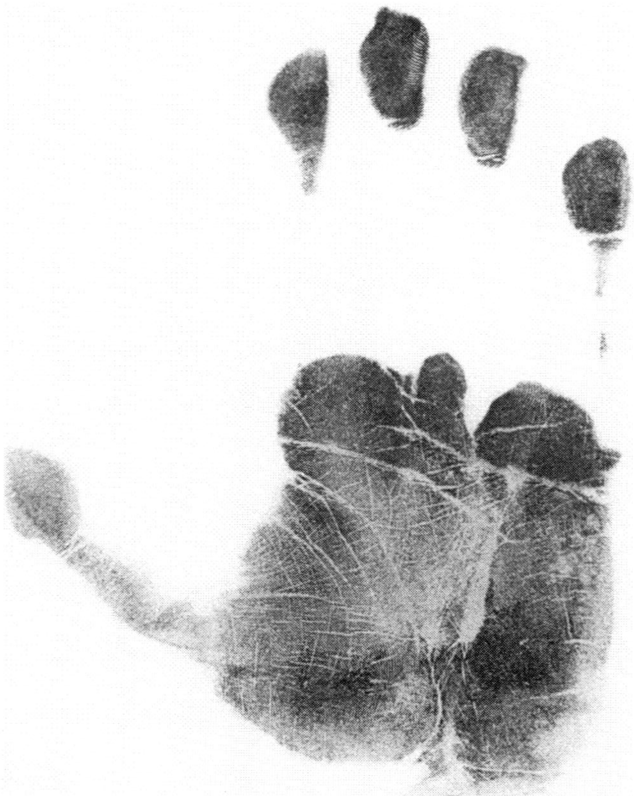

28 Hand-print of a female idiot, showing ankylosis of all fingers

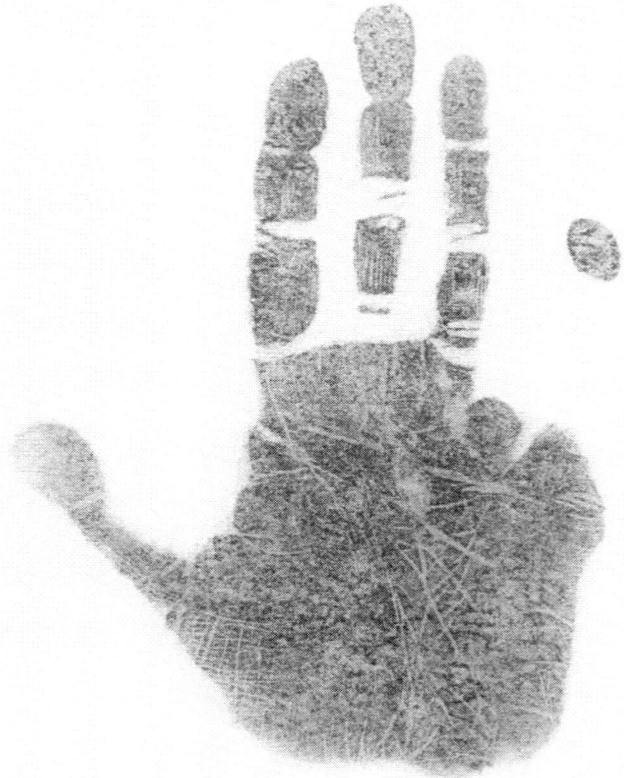

29 Hand-print of a low grade imbecile.
Note the simian-like line, the simian thumb, and the ankylosis of the 5th finger

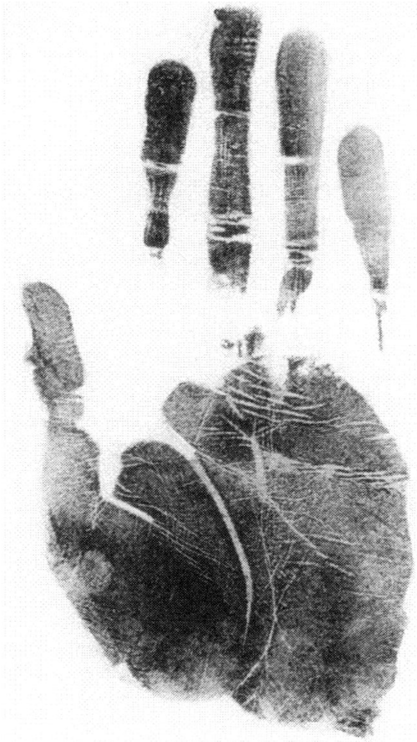

30 Hand-print showing the five-fold semi-circular lines round the 3rd and 4th fingers in an idiot with Bourneville disease.

This is also a case of acromicry

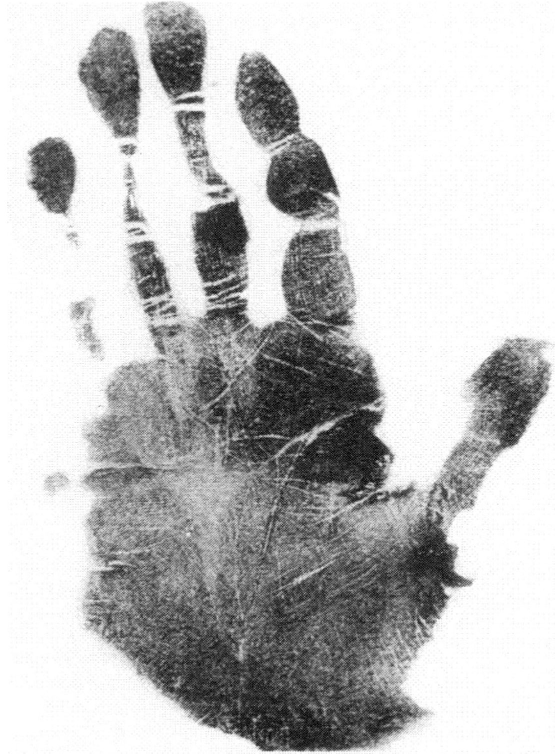

31 A case of acromicry in a mentally defective woman of 70 years of age

HANDS OF HIGH GRADE MENTALLY DEFECTIVE AND OF NORMAL CHILDREN

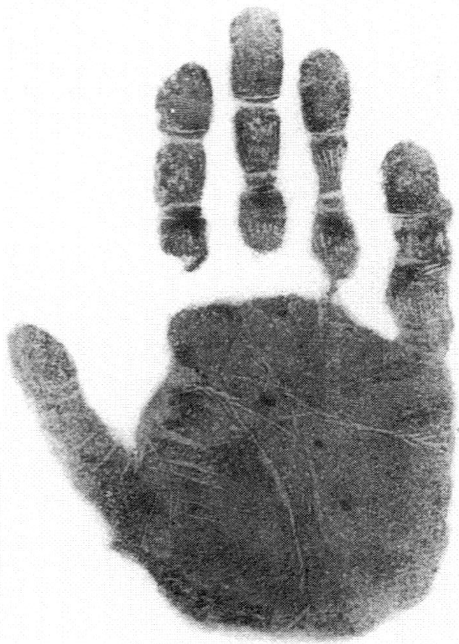

32 Hand-print of a high grade boy of 14 years of age.

Elementary regressive broad hand with short fingers

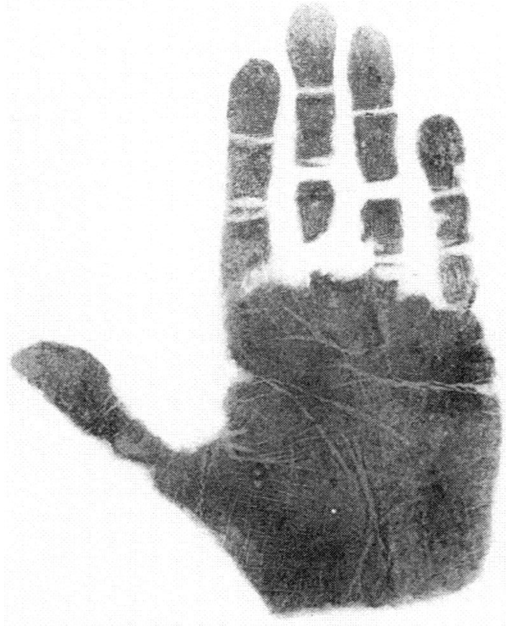

33 Hand-print of a high grade boy of 16 years of age.

Elementary regressive rather narrow hand with long fingers, most typical of that group of boys

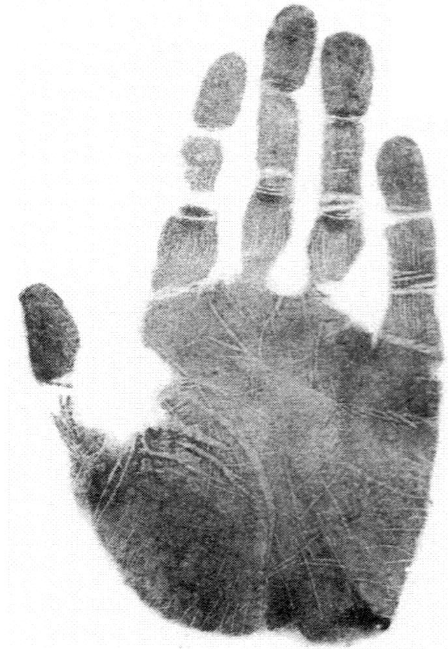

34 Hand-print of a normal boy

HANDS OF HIGH GRADE MENTALLY DEFECTIVE AND OF NORMAL CHILDREN

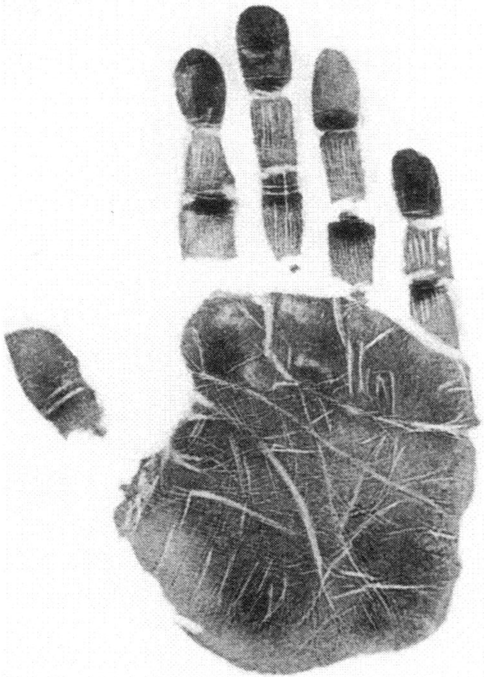

35 Hand-print of a high grade defective girl of 15 years of age – sensitive small type of hand most typical of this group of girls.

Note the simian-like, inarticulate thumb and the very small 5th finger

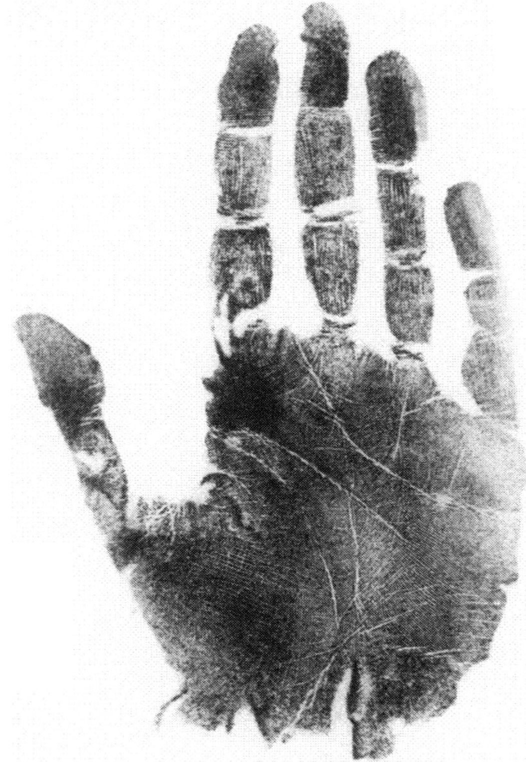

36 Hand-print of a normal girl of 16 years of age

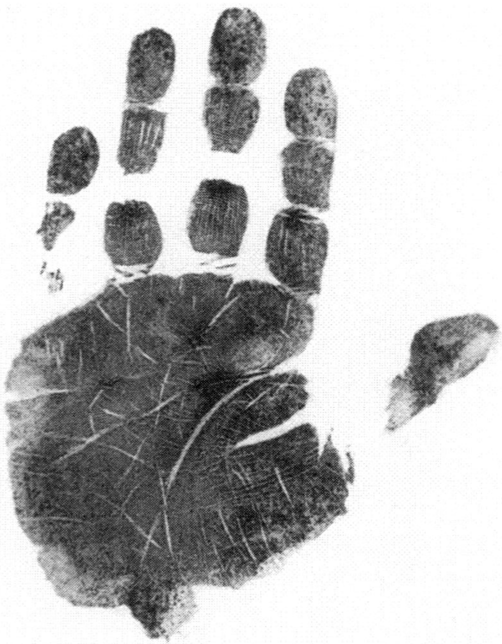

37 Hand-print of an intelligent boy of $3\frac{1}{2}$ years of age, with slight hypopituitarism (obesity).

He is very bright and has a placid and delightful personality

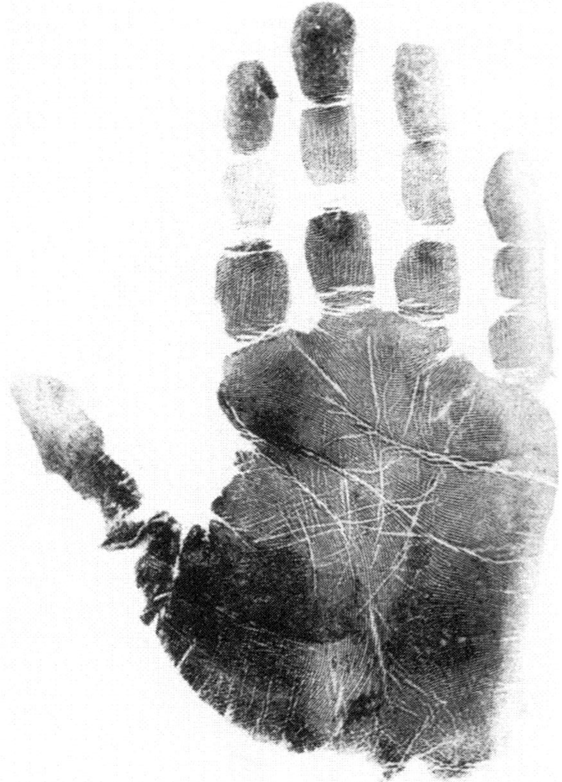

38 Hand-print of a boy prodigy, 8 years of age, precocious in intelligence and artistic development

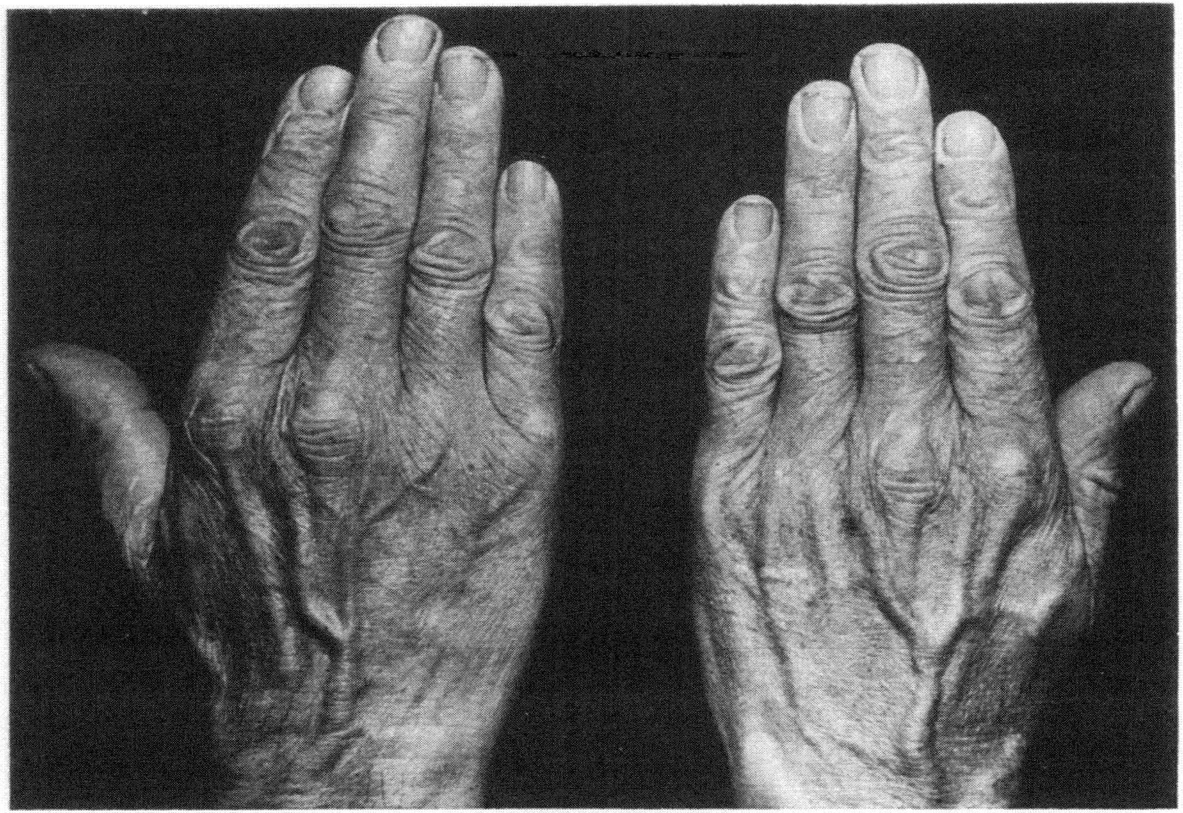

1 Elementary simple hand

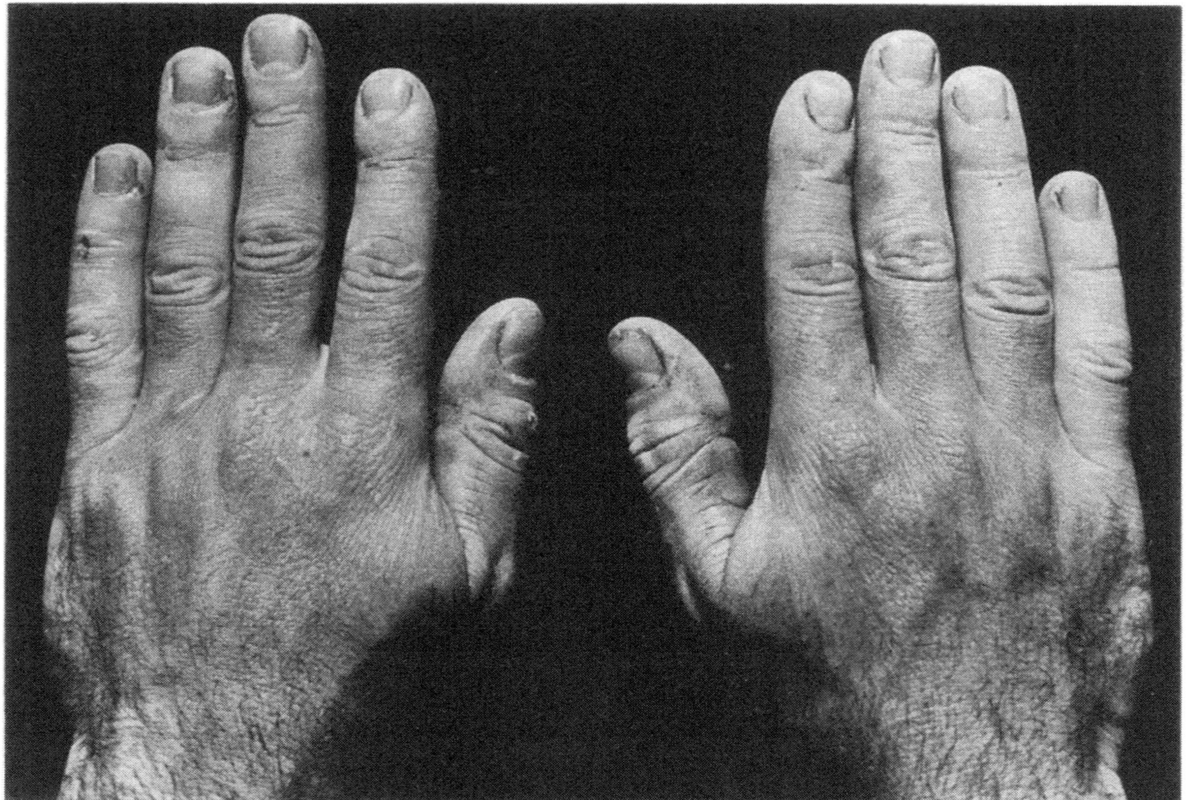

2 Elementary regressive hand

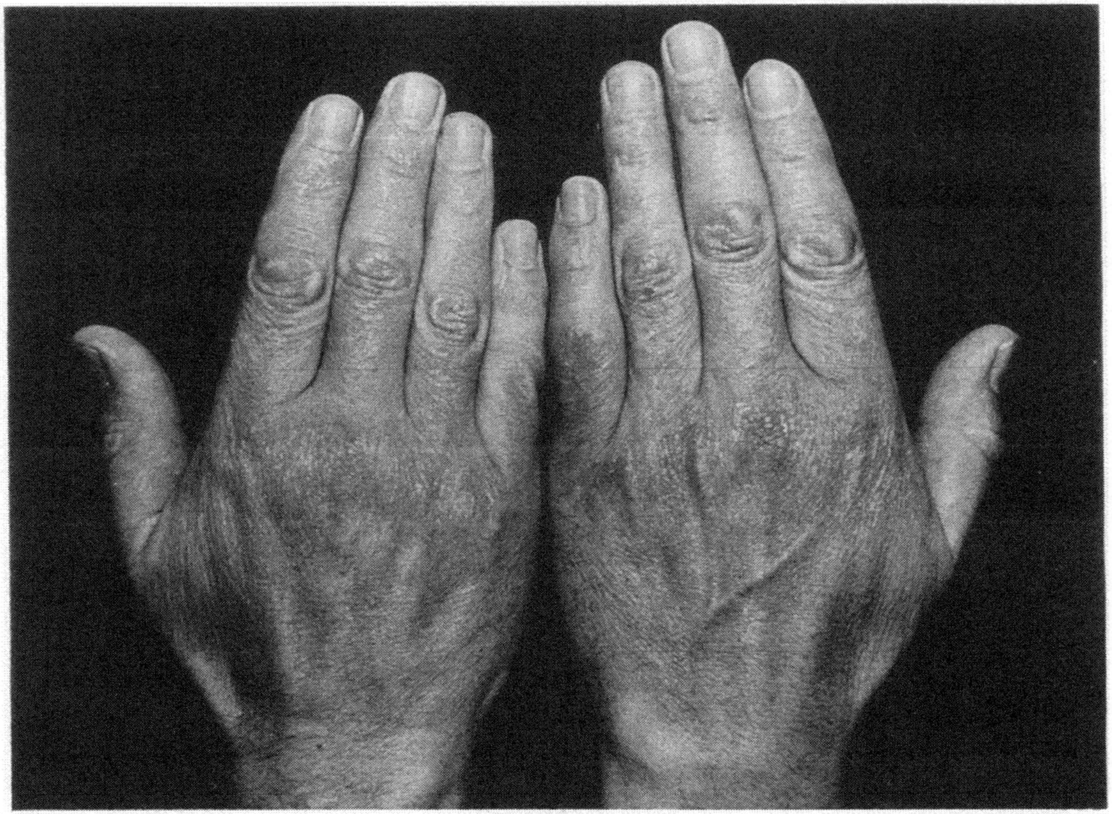

3 Fleshy motor hand

Note the difference in length of right hand and left. The fingers of the left hand are not true to type, and show a " sensitive " factor

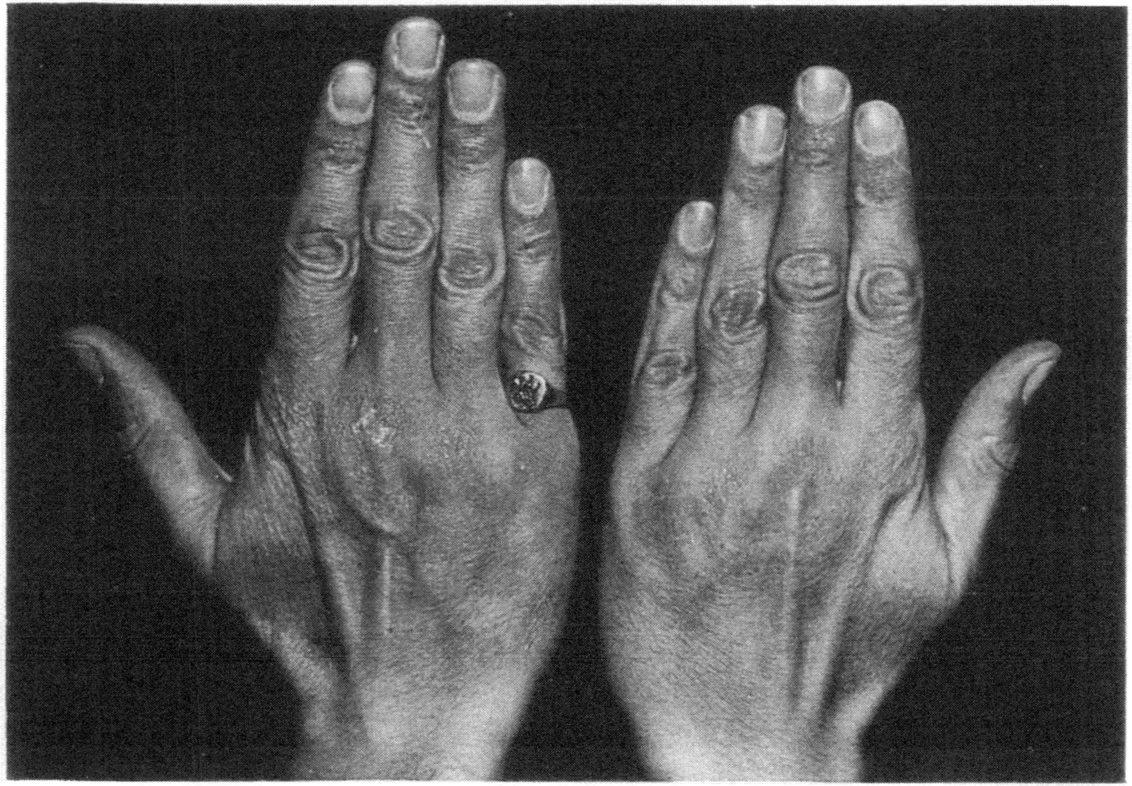

4 Bony motor hand

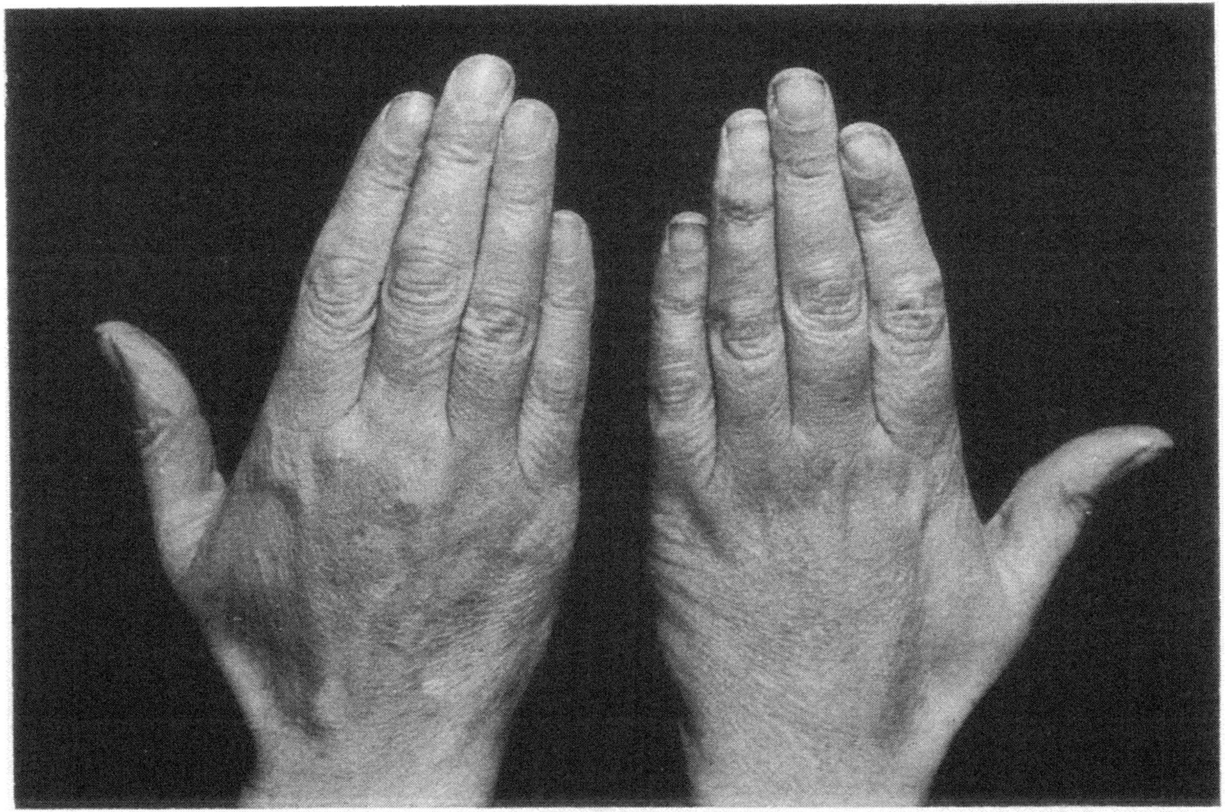

5 Sensitive small hand

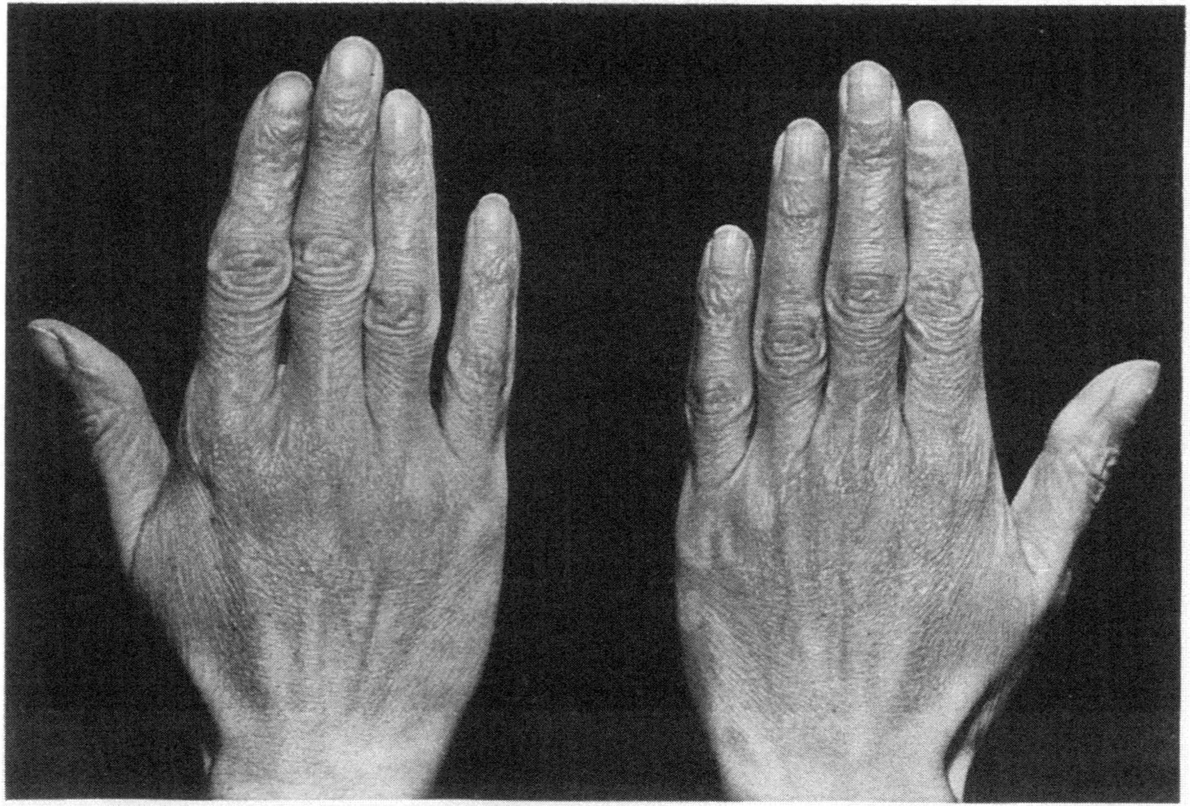

6 Sensitive long hand

BIBLIOGRAPHY

ALLPORT, G. W. *Personality, Psychological Interpretation*. Constable, London, 1938.

—— and VERNON, P. H. *Studies in Expressive Movements*. Macmillan, New York, 1933.

ARISTOTLE. *Des Parties des Animaux*, t. II, livres IV and X.

—— *Histoire des Animaux*, livre I, chap. XII. Translated by Barth. Saint Hilaire.

BELL, Sir Charles. *The Hand, Its Mechanism and Vital Endowments, an Evincing Design*. 5th edition, Murray, London, 1852.

BENDA, C. E. Studies in Mongolism. I. Growth and Physical Development. *Arch. Neurol. Psychiat.*, **41**, pp. 83–97. 1939.

BERMAN, L. *The Glands regulating Personality*. Macmillan, New York, 1921.

BROCA, Paul. L'Ordre des Primates.—Parallele anatomique de l'Homme et des Singes. *Bull. Soc. Anthrop.*, pp. 228–401. Paris, 1869.

BURT, Sir Cyril. *The Measurements of Mental Capacities*. Henderson Trust Lectures, No. VII. Edinburgh, 1927.

CAMERON, N. *The Functional Psychoses. Personality and the Behaviour Disorders*. Vol. II, chap. 29. The Ronald Press Company, New York, 1944.

CARUS, C. G. Ueber Grund und Bedeutung der Verschiedenen Formen der Hand, in *Verschiedenen Personen*. Stuttgart, 1848.

—— *Die Symbolik der Menschlichen Gestalt. Ein Handbuch zur Menschenkenntnis*. Leipzig, 1853.

CHAMBERS, E. G. Statistics in Psychology and the Limitations of the Test Methods. *Brit. Journ. Psychol.*, 1943, vol. XXXIII.

COLLINS and DREVER. *Experimental Psychology*. 6th edition. Methuen, London, 1941.

CROOKSHANK, F. C. *The Mongol in our Midst*, pp. 205–27. Kegan Paul, Trench, Trubner. London, 1931.

CUMMINGS, H. Dermatoglyphics in Indians of Southern Mexico and Central America. *The American Journ. of Phys. Anthrop.*, vol. XV, no. I, Sept.–Dec. 1930.

—— Dermatoglyphic Stigmata in Mongoloid Imbeciles. *The Anatom. Rec.*, **73**, no. 4, April 1939.

—— and SHIRLEY, SPRAGG, D. S. Dermatoglyphics in the Chimpanzee. Description and Comparison with Man. *Human Biol.*, Dec. 1938.

—— and WAITS, W. J., and MCQUITTY, J. T. The Breadths of Epidermal Ridges on the Finger Tips and Palms: A Study of Variation. *The American Journ. of Anat.*, **68**, no. I, Jan. 1941.

CUMMINGS, H. and WRIGHT KENNEDY, Rebecca. Purkinje's Observations (1823) on Finger Prints and Other Skin Features. *The American Journ. of Police Sci.*, vol. XXXI, no. 3, Sept. Oct. 1940.
DUNBAR, H. *Emotions and Bodily Change: a Survey of Literature Psychosomatic Relationships*, 2nd ed. Columbia University Press. 1938.
EARL, C. J. C. The Primitive Catatonic Psychosis of Idiots. *Brit. Journ. Med. Psychol.*, 1934.
FAULDS, H. On the Skin Furrows of the Hand. *Nature*, XXII, p. 605.
FÉRÉ, Ch. Notes sur les Plis de Flexion de la Paume de la Main. *C.R. Soc. Biol.*, f. 52, Paris, 1900.
FLIESS, R. Quoted by Ludwig, W. in *Das Rechts-Links Problem im Tierreich und biem Menschen*. 1932.
FREUDENBERG, W. Zur Frage der Rechtshaendigkeit des Menschen und Gliedmaassenasymetrie. *Zeitsch. f. Säuegetiere*, pp. 36–46. Berlin, 1929.
FRIEDEMANN, A. Handbau und Psychosis. *Arch. f. Neurologie und Psychiatrie*. 1928.
GESELL, A. *The First Five Years of Life*. Methuen, London, 1942.
GURWITCH, A. 'Le Fonctionnement de l'Organisme' d'après K. Goldstein. *Journ. de Psychol.*, Paris, Janvier-Mars, 1939.
HENKEL, K. O. Beitrage zur Entwicklung der Primatenhand. *Morph. Jahrb. Exi.*, pp. 43–8.
HEPBURN, D. The Papillary Ridges on the Hands and Feet of Monkeys and Men. *Sci. Trans. Royal Dublin Soc.*, V, p. 525.
JAMES, W. *Précis de la Psychologie*, 8th edition. Rivière, Paris, 1929.
JUNG, C. G. *Psychologische Typen*. Zuerich and Stuttgart. 1920.
KANNER, L. Behaviour Disorders in Childhood. *Personality and Behaviour Disorders*. Edited by J. McV. Hunt, vol. II, chap. 25. The Ronald Press Company, New York, 1944.
KOLLMAN, A. *Der Tastapparat der Hand*. Hamburg and Leipzig, 1883.
KRAEPELIN, E. *Lehrbych der Psychiatrie*, 8 Ausgabe. Barth, Leipzig, 1915.
KRETSCHMER, E. *Koerperbau und Charackter*, pp. 21–26 and 84. Berlin, 9 und 10 Auflage, 1931.
LECHE, S. M. Handedness and Bimanual Dermatoglyphic Differences. *Amer. J. Anat.*, **53**, pp. 1–53, 1933.
LUDWIG, W. *Das Rechts-Links Problem im Tierreich und biem Menschen*, pp. 423–9. 1932.
MIDLO, Ch. and CUMMINGS, H. Palmar and Plantar Dermatoglyphics in Primates. *The American Anatomical Memoirs*, no. 20. Philadelphia, 1942.
NOGES, A. P. *Modern Clinical Psychiatry*. Philadelphia and London, 1940.
NOIRE. Quoted by OSSERETSKY in 'Psychomotorik', p. 11, *Beih. z. angew. Psychol.*, **57**, 1931.
OSSERETSKY, N. Psychomotorik. *Beihaefte z. Zeitschrift fuer angewandte Psychologie*. Bart, Leipzig, no. 57, 1931.

PARDO-CASTELLO, V. *Diseases of the Nails*. Baillière, Tindall & Cox, London, 1936.

PENROSE, L. S. The Creases on the Minimal Digit in Mongolism. *Lancet*, Sept. 1931.

—— A Clinical and Genetic Study of 7,280 Cases of Mental Defectives. *M.R.C. Special Report Series*. 1938.

—— Auxiliary Genes for determining Sex as Contributory Causes of Mental Illness. *J. Ment. Sci.*, **88**, pp. 308–16. 1942.

—— '*Heredity*' *in Personality and the Behaviour Disorders*, vol. 1, chap. 16. The Ronald Press Company, New York, 1944.

RIFE, D. C. Handedness with special reference to twins. *Genetics*, **25**, pp. 178–186, March 1940.

—— Handedness and Dermatoglyphics in Twins. *Human Biology*, **15**, no. 1, Feb. 1943.

—— Palm Patterns and Handedness. *Science*, no. **94**, 2434, p. 187, Aug. 1941.

—— and CROMWELL, H. Dermatoglyphics in relation to functional handedness. *Human Biology*, **14**, no. 4, Dec. 1942.

ROHRSCHACH, H. *Psychodiagnostics*. Translation and English Edition by Paul Lemkau and Bernhard Kroneneberg. Verlag, Hans Huber, Berne, Switzerland, 1942.

—— *Le Diagnostic Individuel chez l'Enfant au Moyen du Test de Rohrschach*, par M. Loosli-Usteri, Hermann et Cie., Paris, 1938. Actualités Scient. et Industr., no. 639.

ROLLESTON, Sir Humphrey Davy. *The Endocrine Glands in Health and Disease*, pp. 95 and 488. Oxford University Press, 1936.

ROUDINESCO, Mme. et THYSS, J. L'Enfant Gaucher. *Enfance, Revue bimestruelle*, Paris, 1948.

SCHLEGEL, L. Ueber Konstitutionsbiologische Beobachtungen and den Haenden psyschisch abnormer Persoenlichkeiten. Schweizer Arch. f. Neurologie u. Psychiatrie. Band LXI, Zuerich, 1948.

SHELDON, W. H. Constitutional Factors in Personality, in *Personality and Behaviour Disorders*, vol 1, chap. 17. The Ronald Press Company, 1944.

—— and STEVENS, S. S. *The Varieties of Temperament*. Harper, New York, 1942.

—— STEVENS, S. S., and TUCKER, W. B. *The Varieties of Human Physique*. Harper, New York, 1940.

SIMPSON, S. L. *Major Endocrine Disorders*. Kegan Paul, London, 2nd edition, 1948.

STEWART, R. M. A Case of Arachnodactyly. *Arch of Disease in Childhood*, **14**, no. 77, 1939.

STOTT, D. H. *Delinquency and Human Nature*. Carnegie United Kingdom Trust. Dunfermline, 1950.

VALENTINE, C. W. Trends, Dangers and Possibilities in Psychology. *Brit. J. Educ. Psychol.*, 1948.
VASCHIDE, N. *Essai sur la Psychologie de la Main.* Rivière, Paris, 1909.
VIOLA, NACCARATI, and DE GIOVANNI. The Morphological Aspects of Intelligence. *Arch. Psych.*, no. 45.
WALLON, H. Comment se développe chez l'Enfant la Notion du Corps Propre. *J. de Psychol.*, vol. XXVIII. Paris, 1931.
—— *L'Enfant Turbulent.* Alcan, Paris, 1925.
—— *De l'Acte à la Pensée. Essai de psychologie comparée.* Flammarion, Bibliothèque de Philosophie Scientifique. Paris, 1942.
—— *Les Origines de la Pensée chez l'Enfant.* Bibliothèque de Philosophie contemporaine, Presses Universitaires de France, 1945.
—— Le Rôle de 'l'Autre' dans la conscience du 'Moi'. *The Egyptian Journal of Psychology*, **2**, no. 1, 1946.
WERNER, A. *Endocrinology.* Henry Kimpton, London, 1937.
WILSON, E. *Mental Abnormality as related to Hand-Markings.* M.A. Thesis Columbia University, New York, 1924.
WOLFF, Charlotte. The Form and Dermatoglyphics of the Hands and Feet of Certain Anthropoid Apes. *Proc. Zool. Soc.*, London, 1937.
—— A Comparative Study of the Form and Dermatoglyphics of the Extremities of Primates. *Proc. Zool. Soc.*, London, 1938.
—— Character and Mentality as related to Hand Markings. *Brit. J. Med. Psychol.*, 1941.
—— and ROLLIN, H. R. The Hands of Mongolian Imbeciles in relation to their Personality Groups. *J. of Ment. Sc.*, 1942.
—— *The Human Hand.* Methuen, London, 1942.
—— The Hand of the Mental Defective. *British J. Med. Psychol.*, 1944.
—— *A Psychology of Gesture.* Methuen, 1945.
—— The Form and Dermatoglyphics of the Hands of 115 Difficult and High-Grade Boys. *Brit. J. Med. Psychol.*, 1947.
—— La Main et les Tendances Psycho-Biologiques chez l'Enfant. *Enfance, Revue Psychologique*, Paris. Novembre-Dec. 1948.
—— Le Facteur Constitutionel de la Déficiance Mentale et Morale d'après la Main. *Enfance, Revue Psychologique.* Paris, Novembre-Dec. 1949.
WOLFF, W. *The Expression of Personality.* Harper, New York and London. 1943.
WOOD JONES, F. *The Principles of Anatomy as seen in the Hand.* Churchill, London, 1920.

INDEX

Printed in Great Britain by Butler & Tanner Ltd., Frome and London

For Product Safety Concerns and Information please contact our EU representative GPSR@taylorandfrancis.com
Taylor & Francis Verlag GmbH, Kaufingerstraße 24, 80331 München, Germany

www.ingramcontent.com/pod-product-compliance
Ingram Content Group UK Ltd.
Pitfield, Milton Keynes, MK11 3LW, UK
UKHW022155190926
475702UK00003B/46
9781138931954